Relocating Popular Music

Pop Music, Culture and Identity

Pop music lasts. A form all too often assumed to be transient, commercial and mass-cultural has proven itself durable, tenacious and continually evolving. As such, it has become a crucial component in defining various forms of identity (individual and collective) as influenced by factors such as nation, class, gender, ethnicity, location/situation, and historical period.

Pop Music, Culture and Identity investigates the implications of this greatly enhanced status. Particular attention will be paid to issues such as the iconography of celebrity, the ever-expanding archive, the nature of the performance-event, the parameters of generational memory, and the impact of new technologies on global marketing. In particular, the series aims to highlight interdisciplinary approaches and incorporate the informed testimony of the fan alongside a challenging diversity of academic methodologies.

M. King Adkins
NEW WAVE
Image Is Everything

Jennifer Otter Bickerdike
FANDOM, IMAGE AND AUTHENTICITY
Joy Devotion and the Second Lives of Kurt Cobain and Ian Curtis

Ewa Mazierska and Georgina Gregory
RELOCATING POPULAR MUSIC

Rosemary Overall
AFFECTIVE INTENSITIES IN EXTREME MUSIC SCENCES
Cases from Australia and Japan

Pop Music, Culture and Identity
Series Standing Order ISBN 978–1–137–03381–9 (hardback)
(outside North America only)

You can receive future titles in this series as they are published by placing a standing order. Please contact your bookseller or, in case of difficulty, write to us at the address below with your name and address, the title of the series and the ISBN quoted above.

Customer Services Department, Macmillan Distribution Ltd, Houndmills, Basingstoke, Hampshire RG21 6XS, England

---

Relocating Popular Music

Edited by

Ewa Mazierska and Georgina Gregory
University of Central Lancashire, UK

palgrave
macmillan

First published 2015 by
PALGRAVE MACMILLAN

Palgrave Macmillan in the UK is an imprint of Macmillan Publishers Limited, registered in England, company number 785998, of Houndmills, Basingstoke, Hampshire RG21 6XS.

Palgrave Macmillan in the US is a division of St Martin's Press LLC, 175 Fifth Avenue, New York, NY 10010.

Palgrave Macmillan is the global academic imprint of the above companies and has companies and representatives throughout the world.

Palgrave® and Macmillan® are registered trademarks in the United States, the United Kingdom, Europe and other countries.

ISBN 978–1–137–46337–1

This book is printed on paper suitable for recycling and made from fully managed and sustained forest sources. Logging, pulping and manufacturing processes are expected to conform to the environmental regulations of the country of origin.

A catalogue record for this book is available from the British Library.

Library of Congress Cataloging-in-Publication Data

Relocating popular music / [edited by] Ewa Mazierska, University of Central Lancashire, UK, and Georgina Gregory, University of Central Lancashire, UK.
pages cm
Includes bibliographical references.
ISBN 978–1–137–46337–1
1. Popular music—History and criticism. 2. Music and globalization.
I. Mazierska, Ewa, editor of compilation. II. Gregory, Georgina, editor of compilation.
ML3470.R45 2015
781.6309—dc23 2015004020

Typeset by MPS Limited, Chennai, India.

Contents

List of Figures

Notes on Contributors

Srđan Atanasovski is a researcher at the Institute of Musicology SASA in Belgrade and specialises in music and nationalism. He is involved in two projects: *City Sonic Ecology: Urban Soundscapes of Bern, Ljubljana, and Belgrade* (led by the University of Bern) and *Figuring Out the Enemy: Re-Imagining Serbian-Albanian Relations* (led by the Institute for Philosophy and Social Theory in Belgrade). He has published his papers in journals *Musicologica Austriaca, Musicology, Musicological Annual, TheMA* and *Studies in Eastern European Cinema.*

Peter Atkinson is Senior Lecturer and Course Leader in Film and Media Studies at the University of Central Lancashire, specialising in popular music and television. He completed a PhD at Lancaster University in 2006 on the subject of the modern mythology of the Beatles and Liverpool at the beginning of the 1960s and the role of broadcasting in the creation of this myth. Among his publications is an article in *Popular Music and Society* (2010) on the role of BBC radio in giving the Beatles a domestic profile in 1963 and a chapter in *Fifty Years with the Beatles* (2011) on the cumulative character of the Liverpool myth which was synthesised between 1958 and 1963.

Katarzyna Chruszczewska is a doctoral candidate at the Faculty of Interdisciplinary Studies 'Artes Liberales' at the University of Warsaw, confronting in her thesis minor writing with the concept of strong writing of Harold Bloom. Her interests revolve around continental philosophy, aesthetics and popular culture. She has published several articles in Polish and English, including '"To Be Saved by Chaos": Emancipation of Self by Mutilation and Perversion. Chuck Palahniuk's *Invisible Monsters* and *Choke*' in *Fearful Symmetries: Representations of Anxiety in Cultural, Literary and Political Discourses* (ed. Leszek Drong, Jacek Mydła, Wydawnictwo Śląskie, 2013) and '"Każde skazanie jest ostateczne" – gnostyckie *heimarmene* w *Ewangelii według Jezusa Chrystusa* José Saramago' in Światy powieści José Saramago (ed. Wojciech Charchalis, 2013).

Sandra D'Angelo is a doctoral candidate at King's College London, writing a thesis on the transnational circular sound in Brazilian and international baile funk. She is also working for the Horniman Museum in London as music consultant and for the Swiss TV, producing a documentary about baile funk music in Rio de Janeiro. As part of her academic research she is producing an album in the Euro baile funk style, which will be presented to the EDM circuit in 2015.

Georgina Gregory is Senior Lecturer in Film and Media Studies at the University of Central Lancashire where she teaches modules on the visual culture of popular music and youth and popular culture. Her research interests include marginalised forms of music, gender and popular music, music and regional identity, dance culture and boy bands. Georgina has published widely on youth and northern popular culture, glam rock and the 'Madchester' music scene. Her books include the monograph *Send in the Clones: A Cultural Study of the Tribute Band* (2012) and *Careers in Media and Film: the Essential Guide* (co-edited, 2008). For the past 15 years Georgina has worked as a music researcher obtaining data for the UK Performing Rights Society and she performs regularly with a Manchester-based gospel group.

Nick Hodgin is a lecturer at Lancaster University. He is the author of *Screening the East: Heimat, Memory and Nostalgia in German Film since 1989* (2011), co-editor of *The GDR Remembered* (2011) as well as numerous articles and essays on contemporary German film, on East German documentary film and the Deep South in Film. In addition to a project on documentary film culture in the Cold War, he is collaborating on a book about misfits in British Cinema and co-editing *Scars and Wounds*, a collection of essays on trauma and film.

Żaneta Jamrozik is a doctoral candidate at the University of Central Lancashire, writing a thesis about Eminem's music videos. She has published several articles in Polish and in English 'How to be an actress (in Poland): The figure of the actress in Wojciech Jerzy Has's *How to Be Loved* (1962)' in *Studies in Eastern European Cinema* (2013) and 'Acting as Value' in *Work in Cinema* (ed. Ewa Mazierska, 2013).

Ewa Mazierska is Professor in Film and Media Studies at the University of Central Lancashire. She has published over 20 monographs and edited collections, mainly on European and transnational cinema, including *From Self-Fulfilment to Survival of the Fittest: Work in European Cinema from the 1960s till Now* (2015), *Postcolonial Approaches to Eastern European Cinema: Representing Neighbours on Screen* (with Eva Näripea and Lars Kristensen, 2014), and *European Cinema and Intertextuality: History, Memory, Politics* (2011). She is also the author of a monograph on Austrian singer and rapper, Falco, *Falco and Beyond: Neo Nothing Post of All* (2014). Mazierska's work has been translated into almost

20 languages. She is principal editor of a Routledge journal, *Studies in Eastern European Cinema*.

Eva Näripea is Director of Film Archives of the National Archives of Estonia and senior researcher at the Estonian Academy of Arts. She is also the co-editor of *Baltic Screen Media Review* (Baltic Film and Media School of Tallinn University). In 2011 she completed her doctoral dissertation *Estonian Cinescapes: Spaces, Places and Sites in Soviet Estonian Cinema (and Beyond)*. She has co-edited *Via Transversa: Lost Cinema of the Former Eastern Bloc* (2008) with Andreas Trossek, and a special issue on Estonian cinema for *Kinokultura: New Russian Cinema* (2010) with Ewa Mazierska and Mari Laaniste. Her most recent cooperation is an edited volume *Postcolonial Approaches to Eastern European Cinema: Portraying Neighbours on Screen* with Ewa Mazierska and Lars Kristensen (2014). Her research interests include spatial representations in Estonian cinema, histories of Eastern European science fiction film, and reflections of neoliberalism in recent Estonian cinema.

Isabel Stoppani de Berrié is a doctoral candidate at Wolfson College, Oxford, writing a thesis on sound in French film during the period of the Nazi Occupation. She previously achieved a Distinction in the M.St. in Musicology at Oxford, following undergraduate studies in Music at Girton College, Cambridge. Her main research interests lie in twentieth century music, particularly in France; she has recently written on spectral music and the music of Iannis Xenakis.

Lyndon C. S. Way is Assistant Professor of Media and Communications at Izmir University of Economics, Izmir, Turkey. He received a PhD from Cardiff University's Journalism, Media and Cultural Studies department. He has been a radio practitioner and musician, playing in a number of post-punk and indie bands in the UK, Cyprus and Canada. He has published articles on music and politics in *Kültür ve İletişim Dergisi* (2014), *Social Semiotics* (2013) and *Multimodal Communication* (2012). He has also published papers on news representations in *Journalism and Discourse Studies* (2014), *Journal of African Media Studies* (2013), *Journalism Practice* (2013), *Social Semiotics* (2011), *CADAAD* (2011) and *Global Media Journal* (2010 and 2012). He is co-editing a book on music and discourse from a social semiotics perspective. He is the review editor for the *Journal of Interdisciplinary Music Studies*, and on the advisory panel for *Journalism and Discourse Studies*.

Introduction: Setting Popular Music in Motion

Ewa Mazierska

The purpose of this collection is to shed the light on the relations between music and place. Of specific interest to its editors and authors is the idea of relocation of popular music, which can mean many different things: transporting music from one place or historical period to another, hybridising it with a different style, placing it in a new context or furnishing it with a new meaning. On many occasions, this involves discussing music in relation to visual media, most importantly music video, but also film, graphic arts, fashion, and architecture. In our introduction we will present this approach against a background of the existing body of research on place and space in humanities at large, and popular music in particular. This steadily growing research can be seen as a reflection of what we propose to label a 'spatial turn in humanities' and a 'spatial turn in the study of popular music'.

From time to space, from place to space

In the last four decades or so, we observe a rapidly growing interest in the study of place and space in relation to humanities. This shift was recognised or anticipated by Michel Foucault, as demonstrated by his writings from the 1960s and 1970s. Interviewed in 1976 by the editors of the journal *Hérodote*, he said

> A critique could be carried out of this devaluation of space that has prevailed for generations. Did it start with Bergson, or before? Space was treated as the dead, the fixed, the undialectical, the immobile. Time, on the contrary, was richness, fecundity, life, dialectic.
>
> For all those who confuse history with the old schemes of evolution, living continuity, organic development, the progress of consciousness

> or the project of existence, the use of spatial terms seems to have the air of anti-history. If one started to talk in terms of space that meant one was hostile to time. (Foucault 1980: 70)

In the latter part of the interview, Foucault maintains, contrary to the suggestions made by his interlocutors, that there is no conflict between space and time as the principal hermeneutic tools in the humanities. One cannot do without the other, as his own studies on the history of medicine, penal systems or education, demonstrate; an opinion with which we agree. From our perspective another important aspect of his discussion is that by looking at spatial relations we discover relations of power. Terms such as 'position', 'field', 'territory', 'displacement', even belong to political and military discourse (ibid.: 68).

In the lecture published in the 1980s as 'Of Other Spaces', although delivered in 1967, Foucault's position appears to be that already by this point space is winning over history in contemporary culture:

> The great obsession of the nineteenth century was, as we know, history: with its themes of development and of suspension, of crisis, and cycle, themes of the ever-accumulating past, with its great preponderance of dead men and the menacing glaciation of the world. The nineteenth century found its essential mythological resources in the second principle of thermodynamics. The present epoch will perhaps be above all the epoch of space. We are in the epoch of simultaneity: we are in the epoch of juxtaposition, the epoch of the near and far, of the side-by-side, of the dispersed. (Foucault 1986: 229)

In the subsequent decades, this view was developed by a number of thinkers, including Deleuze and Guattari, the authors of the concepts of deterritorialisation and reterritorialisation (Deleuze and Guattari 1983). In 2003, Fredric Jameson began his essay 'The End of Temporality', with the question 'After the end of history, what?', to which he answered:

> [T]he spatial alternative. Statistics on the volume of books on space are as alarming as the birthrate of your hereditary enemy. The rise of intellectual stock of architecture accompanied the decline of *belles lettres* like a lengthening shadow ... So the dictum that time was the dominant of the modern (or the modernism) and space of the postmodern means something thematic and empirical all at once: what we do, according to the newspapers and the Amazon statistics, and what we call what we are doing. I don't see how we can avoid

> identifying an epochal change here, and it affects investments (art galleries, building commissions) as much as the more ethereal things also called values. (Jameson 2003: 696)

Jameson rightly identifies the fact that at the end of the twentieth century and beyond talking about time equals talking about the end or at least exhaustion, as exemplified by some influential books on the contemporary concept of time published in the last decades of the twentieth century and the beginning of the twenty-first century: Lutz Niethammer's *Posthistoire: Has History Come to an End?* (1992), Francis Fukuyama's *The End of History and the Last Man* (1992) or Slavoj Žižek's *Living in the End Times* (2011). By contrast, the influential books about the character of places and spaces do not convey such a sense of finality and fatalism. Since the publication of Gaston Bachelard's *The Poetics of Space* in 1958 (1994), place and space come across as subjective and infinitely flexible. In this way, space is represented in many books written in the shadow of Bachelard's study, for example Jonathan Raban's *Soft City* (1975) and Yi-Fu Tuan's *Topophilia* (1974) and *Space and Place: The Perspective of Experience* (1977). I shall also mention here the influential book by Henri Lefebvre *The Production of Space*, published in 1974.

The shifts from time to space and from history to spatial discourses are also explained by such factors as the proliferation of new media, especially the internet, which allows easy access to many distant sites at once, the increased speed of transportation and, connected with it, growth in migration and diaspora. Jameson specifically mentions the end of colonialism, which leads to bringing many people who previously lived, metaphorically speaking, in different time zones, into one place (Jameson 2003: 700–1), although it can be argued, contrary to his claim, that it is the persistence of colonialism that renders space so important. Migration and international terrorism, all involving the movement of people, commodities, and ideas, can all be seen as consequences of colonialism.

Of course, it will require further investigation to establish whether 'space', 'time', 'temporality', 'spatiality', mean the same things when used by different authors, and whether there is agreement among them as to how to measure the domination of one concept over another; for example, whether it suffices, as Jameson suggests, to check statistics on the number of books containing specific terms, as titles are often misleading. However, it is not our aim here to decipher the exact meanings of these terms. What we want to do is only to 'map' (to use a spatial metaphor) a particular trend in the humanities.

To this trend we shall add a growing interest in the 'everyday', which means not only an interest in what ordinary people do, as opposed to those people who hold political power, but the effect of their actions on space, both its physical qualities and its meanings. The two authors most commonly associated with this trend in humanities, the previously mentioned Henri Lefebvre (2005) and Michel de Certeau (1984), whose work can be seen as a continuation of Walter Benjamin's urban thought, underscore the fact that even by such an ordinary activity as walking in the city we transform the space, giving it a new meaning. Space, most importantly urban space, which is also a 'natural' environment of most popular music genres, is a collection or rather a network of meanings given to it by its users. Lefebvre claims that 'space embodies social relationships' (Lefebvre 1991: 27). These trends are reflected in the rapid development of human geography, a discipline which can be seen as a bridge between the humanities and science.

A third trend in humanities, on which our work reflects and to which it contributes, is a shift from perceiving cultural phenomena as pure and fixed to seeing them as hybridised and undertaking transformation. The father of this type of thinking about culture and society is Karl Marx, who in his most famous and accessible work, *The Communist Manifesto*, included the sentence 'All that is solid, melts into air' (Marx and Engels 2008: 38): a claim which is also regarded as a summary of modernism (Berman 1988) and postmodernism (Harvey 1990). For Marx the reason of this melting is capitalism, which is also an important force in the production of popular music. Such reading of Marx is inspirational for post-Marxist and other scholars exploring the issues of fluidity and hybridity, including urban geographers, anthropologists, sociologists or art historians. For example, the study of transnational cinema and production of films overtakes studies of national cinema and film text. We also notice a growing interest in the migration of people as a chief factor influencing cultural production. The interest in mobility, as opposed to stasis, is reflected in the popularity of certain terms, such as 'performance' (often in the context of performing identity), which supplanted 'being'. 'Space' is also now a more popular term than 'place' or at least 'place' cannot be mentioned without evoking 'displacement'. Again, the previously mentioned Lefebvre, de Certeau, as well as Deleuze and Guattari played a major role in gearing humanity towards examining the motion, the road, rather than its destination. Mike Crang in his essay discussing de Certeau's input into human geography argues that this thinker is 'interested in the relationships of place as a fixed position and space as a realm of practices – counterposing the fixity of the map to the practice of travelling' (Crang 2000: 137–38). Openly or tacitly, the authors of this collection take their cue from de Certeau and privilege space over place, travelling over the fixity of the map. By the same token, they are also more interested in what music does to places and people than what music is.

The spatial turn in the study of music and the role of music in human geography

The situation sketched by Foucault and Jameson also applies to the studies of popular music, as proved by the fast growing number of volumes and academic articles devoted to this subject (for example Bandt et al. 2009; Bennett 2000; Connell and Gibson 2003; Gibson and Connell 2005; Hesmondhalgh and Negus 2002: 205–64; Johansson and Bell 2009; Kun 2005; Lipsitz 1994; Mitchell 1996; Whiteley et al. 2004; Wissmann 2014; Wood et al. 2007), although their authors typically begin their investigation with complaining that there are few previous studies on the interface between music and place.[1] According to David Hesmondhalgh and Keith Negus, place has gained in importance in the studies of popular music roughly from the 1990s (Hesmondhalgh and Negus 2002: 208).

The reason why examining music through the prism of space is a relatively new pursuit is that for a long time music was seen as the ultimate temporal medium. This is because, unlike a painting or a building, music has no 'body', no spatial dimensions, but needs only time to exist (on this argument see Frith 1996: 114–19). In the opinion of many authors, music, at least instrumental, which dominated classical music until Romanticism, does not play a representational function. Not surprisingly, the temporality of music has been the primary object of musicology. This discipline, historically focused on Western classical music, treated this music as universal, as opposed to being located in a specific place or culture. Only with the development of research on different types of music, namely musics produced outside the Western world and popular, by disciplines such as ethnomusicology and sociology of music, place came into the purview of the researchers. This is because the role of place and the cultural environment appears more important when we study foreign cultures than our own. Economic and cultural factors (all related to place and space, because the flow of capital is a spatial phenomenon as well as temporal) also play a greater role in studies of popular music, due to the former being regarded as commercially produced, with an eye for the needs of different types of audiences in terms of age, gender, but also cultural background than in research on classical music, which is more often seen as being a product of an individual author.

Such interest points to the previously mentioned shift in humanities, such as the rapid growth in human geography and development of what can be termed 'philosophy and poetics of place', with its specific vocabulary and theoretical apparatus. For example, music scholars frequently use terms such as ethnoscapes, mediascapes, technoscapes, finanscapes and ideoscapes (all forms of discursive rather than real space), borrowed from anthropologist Arjun Appandurai (see, for example, Lipsitz 1994: 5) and talk about deterritorialisation and reterritorialisation of music (Negus 1999: 153)

The interest in place and space also reflects some specific factors pertaining to the development of popular music and the discourses it activated. In particular, it is argued that the 1980s closed the era of rock as a hegemonic style within popular music (Frith 1981; Grossberg 1992), which allowed other styles, such as hip hop/rap, techno and reggae, to gain more prominent positions. There is a large literature devoted to the importance of place in the history of these genres, greatly exceeding what was ever written about the connection of rock and pop with place. It has been argued that their birth was a result of the combination of distinct economic, technological, cultural and social factors pertaining to one place only: they could not originate elsewhere. For example, hip hop, which emerged in the 1970s, could only be experienced in the Bronx (Graves 2009: 245) or in Black ghettos and other margins of urban America (Perry 2004: 9–37; Rose 1994: 2). A similar argument is proposed in relation to techno, which arguably is a product of Detroit's deindustrialization (Che 2009: 261). It is also believed that rappers and creators of techno are especially spatially aware; they emphasise their connection with place to a much greater extent than rockers or pop singers. The same can be said about their fans who have a greater awareness of the place of their favourite music than fans of rock and pop (Graves 2009: 252). The emphasis of many authors on differentiating between hip hop as Afrodiasporic music, 'rooted' in specific American localities and rap as a transnational style (on the difference between hip hop and rap see, for example, Graves 2009; Krims 2000: 1–16) testifies to the current sensitivity to the place and space on the part of the researchers. By contrast, there is only one word describing blues, jazz or rock, no matter whether it is 'rooted' or 'uprooted'. Without undermining the argument that the link between certain music genres and phenomena, and places, such as rap and the Afro-American ghetto and techno and post-industrial Detroit are particularly strong, we will suggest that foregrounding the connection between them might also be explained by specific power relations permeating scholarship on popular music, namely the fact that these genres and phenomena were initially more likely to appear as exotic or 'ethnic' to those who create discourses on them than the phenomena more associated with the centre, such as pop and rock.

The spatial turn in the study of popular music can also be attributed to the acknowledgement of the role music plays in such phenomena as forging local and national identities on the one hand and abetting cosmopolitanism, globalisation, and tourism on the other. Barry Truax in his book entitled *Acoustic Communication* proposes the concept of an acoustic community, in which sound and music especially creates a unifying and positive relationship with other people belonging to this community. 'Acoustic cues and signals constantly keep the community in touch with what is going on from day to day within it ... The community is linked and defined by its sounds. To an outsider they may appear exotic or go unnoticed, but to the inhabitants they convey useful information about both individual and community life' (Truax 2001: 66). One example from the relatively recent period when music was employed to forge a community was the use of turbo folk during the period of the Balkan wars to unite the Serbian population around a nationalist project (Gordy 1999). On the other hand, what is specific about certain local and national popular musics is the ambition of their creators not to be perceived as local by, for example, not using their native language, but instead singing in English and erasing from their work any possible references to the local identity. This refers, for example, to Swedish pop (Connell and Gibson 2003: 124–5) and 'Vienna electronica' (Huber 2002).

Bob White gives an introduction to the collection *Music and Globalization* a title 'Rethinking Globalization through Music'. Such a title suggests that not only is music a part of the processes of globalisation, but 'there is something about music – and not just world music – that enhances our understanding of globalization' (White 2012: 1). In this context it is worth quoting Michael Bull who argues that 'the age of mechanical reproduction is characterized by increasing sonic saturation in which urban space, both public and private, is colonized' (Bull 2013: 630).

The relationship between popular music and tourism is also intimate and intricate. Not only is music important in tourist practices, as many chapters in this collection demonstrate, but there is something about it which allows us to understand how tourism works. For example, unlike the novel, whose reading is a solitary pastime demanding sitting at home for many hours, listening to a pop song or watching a music video is short and does not demand this kind of commitment. A pop song and music video thus have to capture the audience's attention quickly; therefore it has to take short cuts, communicate with the audience through familiar images and metaphors. Tourism and pop songs are thus based on similar principles: condensation, beautification, romanticisation, mystification and detextualisation (Wang 2000: 160–71).

Studying popular music from the perspective of place and space requires engaging with a number of issues, such as the movement of music, especially between the centre and the periphery and a dynamic between the global and the local. As these terms will also be extensively used by the authors of this collection, it is worth unpacking them. We shall begin with the movement between the centre and the periphery. Music travels practically as long as it exists, but as many authors notice, some routes are more frequented than others. Philip Bohlman observes that 'even more than language, music is the key to understanding and to the power that will turn initial encounter into prolonged dominance. To music, then, accrues the potential to articulate colonial power, and that potential was never lost on those most eager to colonize and missionize the worlds of the others they encountered' (Bohlman 2003: 46–7). In the twentieth century, music is used less in open acts of colonisation and missionisation (not least because they were replaced by subtler forms of dominance), yet its production and consumption reflects well on the imbalances of power between different regions and countries. It has been widely assumed that the centre of popular music, hence its main 'sender' to the rest of the world, is the Anglo-American world. The central position of the States is attributed to its position as a superpower achieved after the Second World War (Bennett 2000: 53); of England to its being the privileged test market and talent pool (Frith 1991). However, this model is increasingly questioned. Simon Frith as early as in 1991 argued that the role of England has been exaggerated and contemporary popular music flows in more directions and more unexpectedly than in the past, not only from the centre to the periphery, but from the periphery to the centre and suggested that this relocation would gain in speed in the following decades (Frith 1991; see also Straw 1991). Of course, it is difficult to measure the extent to which English or Anglo-American music was supplanted by music of different origin, but the successes of such acts as the song *Gangnam Style*, released in 2012 by the Korean singer Psy, subsequently the 'king of YouTube', might suggest that Frith was right. Hence, the 'Anglo-American centre' means at best a historical concept, or at worst an empty signifier, describing a certain comforting, (post)imperial myth of English cultural superiority. That said, the success of *Gangnam Style* might be seen as much as the proof of de-Westernisation of popular music, as of Westernisation of the popular music produced on the periphery, given that the song engages in a parodic way with the Western style, adopted by affluent inhabitants of Korea and its success was ensured by YouTube, a typical Western and capitalist institution. Psy's success might be seen as confirmation of an argument, persuasively presented by Motti Regev, that even if the days of the dominance of Western rock are over, 'rock/pop music', as a certain aesthetics and mode of production still dominates, because they are applied to the emerging phenomena in popular music, such as rap or techno (Regev 2002).

Let us move now to the next contrast: the local and the global. Jan Fairley begins an essay on this subject by provocatively suggesting that what we might see as local pertains to a larger environment or is even global:

> By 'local' do we mean musicians performing in a community space, be it home, tavern or pub, to a local audience? Or do we mean 'local' musicians like the Taraf de Haidouks, a village band from Romania, having their music made available throughout Europe and beyond, wherever it can be distributed, by a small independent company, Melodie? ... Or do we mean by 'local' multinationals investing in local talent so as to thoroughly exploit 'local' markets? (Fairley 2001: 272)

Fairley continues by arguing that global and local suggest a different perspective on the same process. From one perspective 'globalisation' is used to describe the process in which local musicians lose their local identities as they begin to employ musical elements from the global soundscape. From another it refers to the way in which global musicians adopt local sounds (ibid.: 273). Seeing a given phenomenon as local or global also depends on the position adopted by the researchers. Those occupying or dealing with what is widely seen as the 'centre', from the perspective of popular music, namely the Anglo-American world, tend to see as local practically everything that is outside this world; hence the title of Tony Mitchell's book: *Popular Music and Local Identity: Rock, Pop and Rap in Europe and Oceania* (Mitchell 1996). However, the author writing a book about popular music in one (peripheral) country might locate the centre of popular music in its capital, and the local in its provinces.

Fairley and others, suggest that when talking about the local and the global in popular music, we should specify which aspect or stage of the music's life cycle we have in mind: its production, text, or consumption. As I already mentioned, it is widely believed that certain music genres and phenomena are products of a specific geographic and cultural environment. At the same time, their existence and development is possible because of the global flow of music, for example the transportation of recorded music and musical instruments from one part of the world to another. Furthermore, local music might reach global popularity and often this happens because it is seen as being rooted in a specific locality and exotic.

The concept of 'world music' demonstrates well the complementary tendencies of rendering much of popular music as simultaneously local and global, central, and marginal. This term, in common usage, does not designate the totality of the popular music produced in the world, similarly as the term 'world cinema' does not refer to the global film production. Instead, both terms refer to artistic production created outside the Western world, at what can be described as the 'global periphery'. As Jocelyne Guilbault claims, 'world musics are taken to be those musics which come from outside the "normal" Anglo-American (including Canadian and Australian) sources, and mainly from tropical countries' and 'is the product of aggrieved populations, either from third world countries (Africa and the African diaspora) or from disadvantaged population groups in a general sense' (Guilbault 2001: 176). Furthermore, world music, like 'world cinema' needs to be accessible to the Western audience, which means subjected to the same processes of standardisation as the music produced at the centre. Hence the often condescending attitude to world music by some of the audiences, who see it as a product geared towards tourists, as opposed to the 'authentic stuff', which is less accessible, even hidden from Western eyes (on different connotations of world music see also Bohlman 2002; Connell and Gibson 2003: 144–59; Stokes 2003). The term 'world music' thus points to an asymmetry of power within the world of popular music. This does not mean accepting the cultural imperialism thesis in its simplistic form (which appears these days to be completely out of fashion in studies of popular music), according to which a specific empire (most commonly the United States) rules the world of music, but conceding that the world of popular music is postcolonial, which means still largely colonial (Loomba 1998: 12).

The power is more diffuse and difficult to capture today than it was in the first decades after the Second World War, as Frith notices (Frith 1991), because the dominant form of globalisation is glocalisation. It can be argued that the empire of today's cultural imperialism has less in common with a specific state, such as the United States and more with the 'Empire' as discussed by Michael Hardt and Antonio Negri's celebrated book, *Empire* (Hardt and Negri 2000): it is pan-national entity which signifies the power of neoliberal capital, epitomised by YouTube and Google. I shall also add that colonial relations tend to be complex and layered, usually involving more than one set of oppressors and victims. Moreover, colonisation is usually accompanied by self-colonisation, consisting of, for example, imitating the coloniser or integrating elements of colonial culture into one's own cultural production (Bhabha 1994: 85–94). Such behaviour often results from the value of the coloniser's culture in a fight with the oppressors closer to home, such as authoritarian political regimes (Mazierska et al. 2014). This point takes us to the first main subject covered in this collection.

Soundscapes of power

While our collection follows in the footsteps of the existing studies on music and place, it differs from the majority of them in being interested in the way musical works, styles, and their meanings are relocated due to such factors as specific political trends and events, and how places acquire specific meanings through creating, transmitting. and consuming music. Our assumption here is that popular music is in a sense always in a state of relocation and relocates individuals, communities, and places. This also refers to communities created around specific music phenomena; they are not stable, but fragmented and on the move. Hence, musical roots interest us less than *uprooting*, or seeing music allegedly rooted in a specific place and community as belonging to more than one place and community. Similarly, we see place as always being re-worked and changed by music. In this pursuit we follow Will Straw who in an article published in 1991 noticed:

> The long-standing concern of popular-music scholars with the disruption and fragmentation of cultural communities has often masked – in part through nobility of purpose – the investment in imaginary community which underlies it. Those encountering ethnomusicological studies for the first time ... may be struck by the

> prominence within them on notions of cultural totality or claims asserting the expressive unity of musical practices. (Straw 1991: 369)

Straw's suggestion to see musics and places as always relocated was taken, for example, by Patricia Román-Velázquez, who treats salsa as multi-rooted and belonging to more than one place simultaneously (Román-Velázquez 2002: 210). Such an approach inevitably involves approaching popular music through the lens of postmodernism, with its rejection or at least distrust towards concepts such as originality, authenticity, authorship, hierarchy, essence, root and depth, in favour of performativity, anarchy, simulacra, hybridity, dispersal, rhizome and surface (Eagleton 1988; Harvey 1990: 43).

Relocating Popular Music focuses on two types (although connected) of relocation. One consists of taking something from a music or culture located in one place to a music emerging in a new place or a different point in history, where it undergoes profound transformation: being hybridised with a local style, performed in a new location, or listened to by a new type of audience who 'colour' the original music with their own culture. The second type of relocation consists of music affecting the meaning of specific places, in reality or as represented in the visual media, such as music videos and music films. The chapters grouped in the first part show how music produced in what can be seen as the margins of popular music, namely outside the Anglo-Saxon world, feeds on the music produced at the centre, but also transforms it to such an extent that it might be used either as a means to resist the influences from the centre or play a role in local power struggles. By and large, these chapters grapple with the idea of colonialism or even cultural imperialism, drawing, either explicitly or implicitly on postcolonial theory such as the work of Paul Gilroy, Edward Said, Homi Bhabha, and Stuart Hall. They also evoke postmodern and poststructuralist thought, exemplified by Jacques Derrida and the historians of popular music close to postmodernism, such as Simon Frith and Martin Stokes.

The first chapter in this part, 'Turkish Popular Music: People, Places and Spaces of Protest' by Lyndon C. S. Way grapples with these problems by examining music created on the peripheries, but drawing on musical traditions developed in the centre. Way examines the phenomenon of 'Özgün' music as a voice of opposition against the growing Islamisation and authoritarianism of the Turkish state. He begins by noticing the liminal position of Turkey, which stands on the periphery of Europe, straddling the East and West geographically, politically and culturally. Economically, Turkey can be described as a country embracing neoliberal capitalism, which includes privatisation and the lifting of restrictions for international economic interests. At the same time, politically, socially, and culturally, the country grows more conservative, embracing Islamic values and distancing itself from Western cultural practices. 'Özgün' music protests against both trends, seeing neoliberalisation in the economy as leading to impoverishment of a large section of the Turkish population, including the Kurdish minority and depriving those who are not Muslim or reject the close link between the state and religion of many freedoms taken for granted in the Western world. By analysing the work of the Marxist-leaning Grup Kızılırmak, paradigmatic for this type of music, Way shows how Western music is relocated to Turkey and hybridised with local influences to create music which acts as a resistance to the conservative and authoritarian culture. Representations of place also play a key role in expressing resistance. Way pays particular attention to a folk song *My Black Eyed Beauty* on Grup Kızılırmak's video, originally performed in the 1960s. This video perfectly reflects how the band combines Western, Turkish, and more local influences, as well as relocating music from the more liberal past to the authoritarian present to convey its political message.

Sandra D'Angelo in the chapter 'Sampling the Sense of Place in Baile Funk Music' also looks at the music phenomenon created on the peripheries, but influenced by music created elsewhere. Baile funk, an electronic music based on sampling, was born in the most economically deprived areas of Brazil: the favelas. It can be seen as an utterly postmodern phenomenon, hybridising multiple influences and mixing music with noise. However, the crux of D'Angelo's argument is that this genre is a way of 'writing back' to the imperial centre. Drawing on postcolonial theory D'Angelo argues that through the affirmation of baile funk internationally (most importantly in Portugal and other parts of Europe), the favela is reclaiming its place in the hegemonic world. She also discusses baile funk as a means of creating a sense of identity for the inhabitants of the favelas, arguing that the music positively reflects on their attachment to the place where they live and has a utopian dimension, bringing hope of a better life. Such a positive connotation contrasts with the way this genre is seen by the mainstream of Brazilian society as antisocial and possessing a dangerously provocative attitude.

Postmodernism and postcolonialism are also lenses used by Katarzyna Chruszczewska in her examination of the South African hip hop band Die Antwoord, in the chapter titled 'Die Antwoord – the Answer to Unspoken Question'. She claims that the possible reason behind the growing popularity of this band is its unique combination of Western and South African aesthetics. What seems to be the most distinctive feature of Die Antwoord is its relocation of hip hop into the South African music genre of *zef* and the use of parody. The aim of both strategies is to disrupt the post-apartheid topography based on categorisation and exclusion and redefine or re-imagine South Africa in the post-apartheid era. Chruszczewska argues that ultimately Die Antwoord embraces a freak identity as a means to escape fixed categories of class and race and destabilise the boundary between kitsch and art, Western and African, as well as dominant and subaltern culture. To demonstrate the postmodern and postcolonial character of the Die Antwoord phenomenon, Chruszczewska pays particular attention to the visual aspect of the band's performance.

Political events, such as wars and revolutions, play a major role in relocating popular music. This was also the case during the dissolution of Yugoslavia and its aftermath. Srđan Atanasovski in the chapter 'Recycled Music for Banal Nation: The Case of Serbia 1999–2010', addresses the ways in which popular music genres have been recycled in Serbian post-socialist political landscape, focusing on the way the Western-styled music production was relocated, spatially, temporally, and in its meaning, from being a vehicle of the purported 'freedom of expression' in the socialist Yugoslav context, to operating as a mechanism of Serbian banal nationalism. Although the aggressiveness of Serbian nationalism was reflected in various popular music genres, not least in the infamous turbo folk, Atanasovski argues that the state apparatus of both Milošević regime and post-Milošević Serbia endeavoured to banalise the 'hot' Serbian nationalism, striving to represent it in a different, Western-mannered light. Atanasovski shows how the genres such as pop and rock music became involved in the seemingly innocuous representations of Serbian patriotism.

The last chapter in this part, authored by Isabel Stoppani de Berrié, '"Escape and Build Another World": Relocations in Classical Minimalism and Minimal Techno' draws comparison between genres, which belong to different 'meta-genres' of music: popular and classical, and their connections with specific physical spaces. These are classical minimalist music and minimal techno music. Stoppani de Berrié argues that since their inception, these genres bore markers of the spaces in which they first arose, whether in downtown New York or post-Fordist Detroit. She explores how traces of place are retained within later minimal music as it moves into new contexts, comparing reciprocal reworkings of classical and popular minimalist works and foregrounding the role of composers and DJs as relocators. Through dissection of the complex layers of geographical histories present within the new minimalist hybrids more recently performed in urban spaces, she demonstrates that location is more significant than technique in defining differences between minimalist styles. Drawing on a number of case studies she argues that the concept of the minimalist remix as constructing a virtual location harks back to a major trope of early minimal techno in Detroit: that of music signalling escape, through transformation of the urban landscape into a virtual elsewhere. The composer of minimalist music and producer of techno music create their music in a similar way, whether one labels it as 'remixing', 'recomposing', 'retouching', 'reworking', 'reimagining'. Ultimately, Stoppani de Berrié demonstrates that rigid divisions between classical and popular music and by the same token aesthetic hierarchies which such divisions involve are problematic. Her original investigation calls for more studies of this type: namely investigating relocations between popular and classical music and especially those which research such musics in relation to spaces where they are created and performed.

Music, place, and tourism

Music not only reflects and affects the cartographies of power, but also affects the meaning of places, or even, as Simon Frith puts it, defines them (Frith 1996: 125). This is because popular music plays a major role in definition and self-definition of individuals and groups of people (DeNora 2000: 109–31; Whiteley 2004: 4–5). For example, when people think of Brazil, they are most likely to think of samba, connect Argentina with tango and, as was already mentioned, Detroit with techno. Music also helps diasporic communities to preserve their distinctiveness, but also to overcome their isolation and connect with other communities.

However, meaning of any place is never fixed because the factors affecting it are always changing. In this part, we are interested in the way a meaning of a place is transformed by music and in the appropriation and transformation of music work by different agents, including (metaphorical and literal) tourist agents. The relationship between music and tourism is complicated, because tourism is typically linked with superficial or even faked experience, while the Holy Grail of popular music is authenticity. As the authors of a number of chapters included in this part observe, transformation of meaning of a place often happens thanks to genres which combine music with visuals, such as music video, documentary, or scenic film. This is natural as these genres allow us literally to see places, not only imagine them.

This part begins with Peter Atkinson's chapter 'Abbey Road Studios, the Beatles and the Tourist Gaze'. Its author moves away from the dominant representation of the Beatles as linked to Liverpool and focuses on the London chapter in their working life, as well as what can be described as their after-life, hence the Beatles already relocated from their roots. Atkinson uses the concept of the 'tourist gaze', elaborated by John Urry and Jonas Larsen, for two main goals. One of them is comparing the work of the Beatles in different stages of their careers. Atkinson argues that while in the early period Beatles' songs tended to have general subjects, such as love and youth, in the later period they were more specific, focusing on particular locations, and often assumed the position of a detached tourist gazing into the life of the London affluent. Finally, in their later period they often reflected on their extensive travels and assumed the place of meta-tourists, assessing the experience of other tourists and new forms of tourism, emerging at the time. Atkinson's second goal is to examine how the Beatles' use of the Abbey Road Studios for the arguably most mature chapter in their career affected the meaning of this place from the 1960s to the 1990s. He discusses how the studios were transformed into an archive and museum of the Beatles and how this status was furthered by the development of the recording technology, most importantly the invention of the CD. He also accounts for the fact that the Beatles themselves or at least one Beatle, Paul McCartney, took an active role in bestowing on Abbey Road Studios the meaning of a heritage site of great importance for Beatles' history and English culture at large. Ultimately Atkinson argues that there is a two-way relationship between the music and place as an object of tourism. The place can be invested with an aura of the artist connected to the place and the special character of the place can help the artist to prolong his/her career.

Eva Näripea's 'East Meets West: Tallinn Old Town and Soviet Estonian Pop Music on Screen' looks at musical films – both short concert programmes and full-length musicals – produced in Soviet Estonia in the late 1960s and early 1970s. Näripea argues that the entire cycle is characterised by the retro-medieval setting, inspired by Tallinn's medieval Old Town, a picturesque environment that has always been an attractive source of imagery for visual media, especially in connection with the rise and development of modern tourism practices. At the same time, the Old Town has also been a place for negotiations between conflicting ideologies and (national) identities, and an arena of central importance for games of power, resistance, and adaptation. These processes generated a particularly complex configuration of representations and practices during the 1960s and 1970s. Spatially, these entertaining 'sensuous utopias', to draw on a notion by Richard Dyer, follow the logic of John Urry's tourist gaze and are firmly related to the official Soviet discourse that served the ideological and representational needs of the central administration in Moscow. Musically, they reflect the gradual emergence and popularisation of Western forms, such as jazz and pop music. In both dimensions, they bear witness to the battles fought in the cultural forefront of the Cold War where the Soviet youth and popular culture became one of the central targets of the Western ideologues who bombarded their easy prey with products of capitalist entertainment. At the same time, the films, projecting a distinctive image and sound of the Baltic periphery as the 'Soviet West', were also meant to serve as glamorous displays of Soviet 'progressiveness' for audiences on both sides of the Iron Curtain on the one hand and to function as 'ventilation valves', or illusory tours of the imaginary 'West', for those trapped in the confines of the 'one sixth of the world' on the other.

The next chapter, authored by myself, 'Tourism and Heterotopia in Falco's Songs' examines the career and textual characteristics of the songs and videos of Austrian singer and rapper Falco, who gained worldwide popularity in the mid-1980s. Falco can be regarded as both a local and a global pop musician, as he never renounced his native language, German (and on occasions used Viennese dialect), yet usually mixed German with English and other languages. Moreover, he was singing about his native city, Vienna, but also about many exotic locations, such as Constantinople, Kenya, Hawaii, Panama, the Himalayas, Jamaica, Arizona, Toronto, New York, Amman. Travel is a constant theme of his songs and he engages with its different forms, such as tourism, colonial expansion, and time travel. What interests me especially is the transformation of his discourse on tourism, taking place in his first three records, of which the first one, *Einzelhaft* (1982), was addressed primarily to the Austrian audience and the last one, *Falco 3* (1985), to the global audience. My argument is that many of Falco's songs and videos produce an effect of heterotopia, as defined by Michel Foucault: of an impossible space, in which different spaces and ontological orders mingle, but without producing a smooth, coherent space. By the same token, Falco's songs, although superficially presenting the unproblematic pleasures of tourism, in fact produce an uncanny effect that distant places and cultures cannot be reached. Ultimately Falco's songs point to the contemporary experience of time and space as fragmented and pose a question about the purpose and value of cultural relocation.

The last three chapters in this part discuss cases which are close to the phenomenon known as 'dark tourism'. This activity is defined as visiting places of atrocity and suffering, such as murder sites, battlefields, natural and accidental disasters, and cemeteries (Frew and White 2013: 2; Lennon and Foley 2000: 2–4). I have not myself encountered this term in the discourse of popular music, but the associations of other terms, used in the discussion of places where music is produced and consumed, such as the 'hip hop sublime' (Krims 2002), are similar to that of dark tourism. Moreover, it is a well-known fact that sites of celebrity deaths and graves are common destinations of music fans (Best 2013). The authors of the subsequent chapters do not deal precisely with such phenomena, but they engage with the special interest music grants to sites of poverty and misery, often seen as sites of 'authenticity' and the role of music in 'darkening' or 'lightening' a particular landscape. In the chapter entitled 'Emotional Places: Relocating Mancunian Melancholia' Georgina Gregory evokes the concept of Manchester and the North of England as dark and melancholic. She argues that for this reason many observers of the artist Morrissey's identity is specifically connected to the city of Manchester and northern England more generally. Not only was Morrissey born in Manchester and has a local accent, but his song writing is famously morbid and self-flagellating, and full of references to obscure places and people. Moreover, the city is featured in the Smiths' artwork. It is therefore puzzling to find that despite the fact that he is so closely associated with Manchester and could hardly be described as pandering to the specific needs of global audiences, Morrissey has a large and growing fan following in California where he is something of a popular icon among Mexican Americans. To explain it, the author of this chapter reflects more generally on the role played by place in the culture of popular music. This involves establishing how regional identity and references to locality influence the reception of pop songs. She also looks at the relationship between the local and the global from the perspective of those involved in the marketing of popular music. The problems associated with translating popular songs are discussed to show how language can present certain barriers towards international acceptance. She also looks at strategies employed within the music industry to overcome these obstacles.

Nick Hodgin discusses the 2003 documentary film by British director Andrew Douglas, *Searching for the Wrong-eyed Jesus*, in which musician Jim White travels across the rural South of the USA searching for 'the soul of the South'. In reality, the authors of the film, as Hodgin observes, try to establish what it is about the South that has mesmerized artists. He answers that this has to do with the perceived authenticity of this region and especially its music, which is seen as the cultural expression of the poor. Authenticity, as the author observes, has particular currency in the discourse on popular music, which explains why the South is a magnet to musicians, including those who have roots elsewhere. Hodgin argues that the case of the South and its music demonstrates that 'authenticity' is not a reflection of the real state of things, but a construction, whose aim is to add value to music and attract audiences. Hence, authenticity involves 'simulacra', constructing a copy which comes across as more convincing than the original. Of interest to him is thus the postmodern interest in authenticity and simulation, between original and copy, explored by Eco and by Baudrillard, both of whom focus in particular on North American culture as one characterised by a search for the authentic and by the proliferation of copies, fakes and reproductions, which in Eco's words provide 'reassurance through imitation'. Hodgin considers their ideas in the context of music and place and highlights the ascription of authenticity by outsiders, a practice he relates to the essentialising and romanticising tendencies discernible in tourists' and travellers' explorations of other cultures.

A different approach is adopted by Żaneta Jamrozik. Her point of departure is an observation that Detroit, once a model motor city, after the 1967 riot was represented in the media and political discourse as a place of perpetual crisis, a dystopia, being an object of what is labelled 'ruin porn', an aesthetic which can be linked to 'dark tourism'. Both representations, she argues, are only partially true and can be seen as political constructions rendering the constantly redefining itself city as always the same. Jamrozik points to the role of Eminem in opposing this static image of Detroit by shooting video to his song *Beautiful* in the city and she contrasts the video with the Chrysler advert, which, although referring to the video, presents the city as stable, 'recovered dystopia'. She argues that the *Beautiful* video attempts to construct a more fluid image of Detroit not in the sense of offering an ideal and impossible image of this city, but one rooted in the personalised experience of the city, where the past, the present and the future intermingle with each other. In her understanding of the experience of the city, she draws on the thought of Henri Lefebvre and Walter Benjamin, who saw the modern city as layered and eluding a simple dichotomy of good and bad. She also argues that the video to *Beautiful* aligns the personal history of Eminem, marked by numerous personal problems, with the history of his city. In this work, Eminem reinvents (relocates) himself. The video presents Eminem turning to the city space in order to sort out his personal problems and finish the song. Jamrozik also draws attention to the specific affinity between music video and movement, resulting from music video always presenting space in flux.

As was demonstrated by the outline of the content, this collection considers many corners of the world, such as Brazil, South Africa, Turkey, the USA, former Yugoslavia, England, Austria and France. Equally, it examines many genres of popular music, such as rock, pop, techno, jazz and folk. However, rap plays a privileged part in it.[2] Not only are there more chapters on rap than on any other genre, but most authors refer to 'rap aesthetics', consisting of sampling and hybridising different elements of the soundscape, to produce a new soundtrack (Shusterman 1992: 204). This reflects the fact that rap, described as the ultimate postmodern musical style (Potter 1995) (in common with bricolage, which is seen as the privileged postmodern style in plastic arts), provides a blueprint for creating music in the late twentieth and twenty-first centuries, in a large part due to the development of digital technologies. Today it is widely believed that such music is no less original than the earlier styles, but also functions as a sonic archive, helping to understand how music is created and relocated. Such knowledge, in my opinion, does not diminish the pleasure of appreciating music, but adds to it, confirming the view of the philosopher Nelson Goodman, that aesthetic experience is ultimately an epistemological experience: the more we know about music, the more we appreciate it (Goodman 1968: 258–65).

The authors of many chapters included in this collection show particular sensitivity to the issue of relocation due to being themselves relocated through living in a different country from that where they were born. Most of them also perform a kind of virtual tourism by writing about music which is far from their roots. Some, additionally, departed from their core academic interests. This also refers to me, as I am a Pole working in the UK and writing about Austrian music. I also recently relocated my academic interests by embarking on the study of popular music after more than a quarter of century of researching cinema, albeit such transnational phenomena as road cinema, émigré directors and postcolonialism. However, irrespective of whether the authors write about far away places and musics or those closer to home, they attempt to assess the value, as much political as aesthetic, of relocation and in most cases they argue that it plays a positive role: it leads to creating a new music without destroying or diminishing the old one. They agree with Simon Frith, who in relation to rap claims that 'far from musical authority being dissipated into fragments and second-hand sounds it is enhanced by the attention drawn to the quoting act itself' (Frith 1996: 115). They also argue that relocating music sheds a new light on the 'original' or better say 'hypotext', similarly as an act of translation reveals something about the original of which its authors and original readers were not aware, and prolongs its existence. In line with this argument we hope that this book will not only reveal the richness and complexity of popular music, but also add to it.

Notes

1. This in part reflects the fact that, until recently, popular music was not taken seriously and hence its relationship with other areas of humanities is unbalanced. The authors of studies of popular music draw on such disciplines as sociology, psychology, geography, or film studies, but sociologists, psychologists, geographers or film historians rarely repay them with the same interest.
2. Rap is also considered in the vast majority of books dealing with popular music and space, for example the volume edited by Whiteley, Bennett, and Hawkins devotes to its second part to it (Whiteley et al. 2004: 89–146). However, our collection, more than others, draws attention to rap aesthetics as a means of relocation of music.

Works cited

Bachelard, Gaston (1994) [1958]. *The Poetics of Space*, trans. Maria Jolas (Boston: Beacon Press).

Bandt, Ros, Michelle Duffy and Dolly MacKinnon (eds) (2009). *Hearing Places: Sound, Place, Time and Culture* (Newcastle: Cambridge).

Bennett, Andy (2000). *Popular Music and Youth Culture: Music, Identity and Place* (Houndmills, Basingstoke: Palgrave).

Berman, Marshall (1988). *All That Is Solid Melts into Air: The Experience of Modernity* (New York: Penguin).

Best, Gary (2013). 'Dark Detours: Celebrity Car Crash Deaths and Trajectories of Place', in Leanne White and Elspeth Frew (eds), *Dark Tourism and Place Identity: Managing and Interpreting Dark Places* (London: Routledge), pp. 202–16.

Bhabha, Homi K. (1994). *The Location of Culture* (London: Routledge).

Bohlman, Philip V. (2002). *World Music: A Very Short Introduction* (Oxford: Oxford University Press).

Bohlman, Philip V. (2003). 'Music and Culture: Historiographies of Disjuncture', in Martin Clayton, Trevor Herbert and Richard Middleton (eds), *The Critical Study of Music: A Critical Introduction* (London: Routledge), pp. 45–56.

Bull, Michael (2013). 'Remaking the Urban: The Audiovisual Aesthetics of Ipod Use', in John Richardson, Claudia Gorbman and Carol Vernallis (eds), *The*

Oxford Handbook of New Audiovisual Aesthetics (Oxford: Oxford University Press), pp. 628–44.
Che, Deborah (2009). 'Techno: Music and Entrepreneurship in Post-Fordist Detroit', in Ola Johansson and Thomas L. Bell (eds), *Sound, Society and the Geography of Popular Music* (Farnham: Ashgate), pp. 261–80.
Connell, John and Chris Gibson (2003). *Sound Tracks: Popular Music, Identity and Place* (London: Routledge).
Crang, Mike (2000). 'Relics, Places and Unwritten Geographies in the Work of Michel de Certeau (1925–1986)', in Mike Crang and Nigel Thrift (eds), *Thinking Space* (London: Routledge), pp. 136–53.
de Certeau, Michel (1984). *The Practice of Everyday Life*, trans. Steven Rendall (Berkeley: University of California Press).
DeNora, Tia (2000). *Music in Everyday Life* (Cambridge: Cambridge University Press).
Deleuze, Gilles and Félix Guattari ([1972] 1983). *Anti-Oedipus: Capitalism and Schizophrenia*, trans. Robert Hurley, Mark Seem and Helen R. Lane (Minneapolis, MN: University of Minnesota Press).
Eagleton, Terry (1988) [1985]. 'Capitalism, Modernism and Postmodernism', in David Lodge (ed.), *Modern Criticism and Theory: A Reader* (London and New York: Longman), pp. 384–98.
Fairley, Jan (2001). 'The "Local" and "Global" in Popular Music', in Simon Frith, Will Straw and John Street (eds), *The Cambridge Companion to Pop and Rock*, pp. 272–89.
Foucault, Michel (1980). 'Questions of Geography', in his *Power/Knowledge* (New York: Harvester Wheatsheaf), pp. 63–77.
Foucault, Michel (1998) [1986]. 'Of Other Spaces', trans. Jay Miskowiec, in Nicholas Mirzoff (ed.), *The Visual Culture Reader*, 2nd edition (London: Routledge), pp. 229–36.
Frith, Simon (1981). *Sound Effects* (New York: Pantheon).
Frith, Simon (1991). 'Anglo-America and its Discontents', *Cultural Studies*, 3, pp. 263–69.
Frith, Simon (1996). 'Music and Identity', in Stuart Hall and Paul du Gay (eds), *Questions of Cultural Identity* (London: Sage), pp. 108–27.
Frew, Elspeth and Leanne White (2013). 'Exploring Dark Tourism and Place Identity' in Leanne White and Elspeth Frew (eds), *Dark Tourism and Place Identity: Managing and Interpreting Dark Places* (London: Routledge), pp. 1–10.
Fukuyama, Francis (1992). *The End of History and the Last Man* (London: Hamish Hamilton).
Gibson, Chris and John Connell (2005). *Music and Tourism: On the Road Again* (Clevedon: Channel View).
Goodman, Nelson. 1968. *Languages of Art: An Approach to a Theory of Symbols* (Indianapolis: Bobbs-Merrill).
Gordy, Eric D. (1999) *The Culture of Power in Serbia: Nationalism and the Destruction of Alternatives* (University Park, PA: Pennsylvania State University Press).
Graves, Steven (2009). 'Hip Hop: A Postmodern Folk Music', in Ola Johansson and Thomas L. Bell (eds), *Sound, Society and the Geography of Popular Music* (Farnham: Ashgate), pp. 245–60.
Grossberg, Lawrence (1992). *We Gotta Get Out of This Place* (London: Routledge).
Guilbault, Jocelyne (2001). 'World Music', in Simon Frith, Will Straw and John Street (eds), *The Cambridge Companion to Pop and Rock* (Cambridge: Cambridge University Press), pp. 176–92.

Hardt, Michael and Antonio Negri (2000). *Empire* (Cambridge Massachusetts: Harvard University Press).
Harvey, David (1990). *The Condition of Postmodernity* (Oxford: Blackwell).
Hesmondhalgh, David and Keith Negus (2002). 'Editors' Introduction to Section IV' in David Hesmondhalgh and Keith Negus (eds), *Popular Music Studies* (London: Arnold), pp. 207–9.
Huber, Michael (2002). 'The Case Study of Vienna Electronica', in *Creative Europe: On Management and Governance of Artistic Creativity in Europe* (Bonn: ERICarts), pp. 269–80.
Jameson, Fredric (2003). 'The End of Temporality', *Critical Inquiry*, 29, pp. 695–717.
Johansson, Ola and Thomas L. Bell (eds) (2009). *Sound, Society and the Geography of Popular Music* (Farnham: Ashgate).
Krims, Adam (2000). *Rap Music and the Poetics of Identity* (Cambridge: Cambridge University Press).
Krims, Adam (2002). 'The Hip-hop Sublime as a Form of Commodification', in Regula Burckhardt Quereshi (ed.) *Music and Marx* (London: Routledge), pp. 63–78.
Kun, Josh (2005). *Audiotopia: Music, Race, and America* (Berkeley: University of California Press).
Lefebvre, Henri (1991) [1974]. *The Production of Space* (Oxford & Cambridge: Blackwell).
Lefebvre, Henri (2005). *Critique of Everyday Life Volume III: From Modernity to Modernism* (London: Verso).
Lennon, John and Malcolm Foley (2000). *Dark Tourism* (London: Continuum).
Lipsitz, George (1994). *Dangerous Crossroads: Popular Music, Postmodernism and the Poetics of Place* (London: Verso).
Loomba, Ania (1998). *Colonialism/Postcolonialism* (London: Routledge).
Marx, Karl and Friedrich Engels (2008) [1848]. *The Communist Manifesto*, with an introduction by David Harvey (London: Pluto).
Mazierska, Ewa, Lars Kristensen and Eva Näripea (2014). 'Introduction: Postcolonial Theory and the Postcommunist World', in Ewa Mazierska, Lars Kristensen and Eva Näripea (eds), *Postcolonial Approaches to Eastern European Cinema: Portraying Neighbours On-Screen* (London: I. B. Tauris), pp. 1–39.
Mitchell, Tony (1996). *Popular Music and Local Identity: Rock, Pop and Rap in Europe and Oceania* (London and New York: Leicester University Press).
Negus, Keith (1999). *Music Genres and Corporate Cultures* (London: Routledge).
Niethammer, Lutz (1992) [1989]. *Posthistoire: Has History Come to an End?*, trans. Patrick Camiller (London: Verso).
Perry, Imani (2004). *Prophets of the Hood: Politics and Poetics in Hip Hop* (Durham: Duke University Press).
Potter, Russell A. (1995). *Spectacular Vernaculars: Hip-Hop and the Politics of Postmodernism* (Albany, NY: State University of New York Press).
Raban, Jonathan (1975). *Soft City* (Glasgow: Fontana).
Regev, Motti (2002). ''The Pop-Rockization of Popular Music', in David Hesmondhalgh and Keith Negus (eds), *Popular Music Studies* (London: Arnold), pp. 251–64.
Román-Velázquez, Patricia (2002). 'Locating salsa', in David Hesmondhalgh and Keith Negus (eds), *Popular Music Studies* (London: Arnold), pp. 210–22.
Rose, Tricia (1994). *Black Noise: Rap Music and Black Culture in Contemporary America* (Hanover: Wesleyan University Press).

Shusterman, Richard (1992). *Pragmatist Aesthetics: Living Beauty, Rethinking Art* (Oxford: Blackwell).
Stokes, Martin (2003). 'Globalization and the Politics of World Music', in Martin Clayton, Trevor Herbert and Richard Middleton (eds), *The Critical Study of Music: A Critical Introduction* (London: Routledge), pp. 297–308.
Straw, Will (1991). 'Systems of Articulation, Logics of Change: Communities and Scenes in Popular Music', *Cultural Studies*, 3, pp. 368–88.
Truax, Barry (2001). *Acoustic Communication*, 2nd edition (Westport, Connecticut: Ablex).
Tuan, Yi-Fu (1974). *Topophilia: A Study of Environmental Perception, Attitudes, and Values* (New York: Columbia University Press).
Tuan, Yi-Fu (1977). *Space and Place: The Perspective of Experience* (Minneapolis: University of Minnesota Press).
Wang, Ning (2000). *Tourism and Modernity: A Sociological Analysis* (Amsterdam: Pergamom).
Whiteley, Sheila (2004). 'Introduction' to Andy Bennett and Stan Hawkins (eds), *Space and Place: Popular Music and Cultural Identity* (Aldershot: Ashgate), pp. 1–22.
White, Bob W. (ed.) (2012). *Music and Globalization* (Bloomington: Indiana University Press).
Whiteley, Sheila (2004). 'Introduction' to Sheila Whiteley, Andy Bennett and Stan Hawkins (eds), *Space and Place: Popular Music and Cultural Identity* (Farnham: Ashgate), pp. 1–22.
Whiteley, Sheila, Andy Bennett and Stan Hawkins (eds) (2004). *Space and Place: Popular Music and Cultural Identity* (Farnham: Ashgate).
Wissmann, Torsten (2014). *Geographies of Urban Sound* (Aldershot: Ashgate).
Wood, Nicola, Michelle Duffy and Susan Smith (2007). 'The Art of Doing (Geographies of) Music', *Environment and Planning D: Society and Space*, 25, pp. 867–89.
Žižek, Slavoj (2011). *Living in the End Times* (London: Verso).

Part I
Soundscapes of Power

1
Spaces of Protest in Turkish Popular Music

Lyndon C. S. Way

Turkey stands on the periphery of the dominating centre of Europe, straddling the East and West politically, culturally, and geographically. Oppositional Turkish voices come from the periphery of Turkey itself, challenging the government's political Islamist centre. 'Özgün' (authentic protest) music is one such voice of protest. With its roots in Anatolian musical traditions, it also borrows liberally from Western musical styles, instrumentation and even harmonies raising fears of cultural imperialism where the West is seen to dominate the rest. This chapter argues that it is useful to think of this borrowing in terms of relocating semiotic resources from both the West and Turkish culture to construct spaces of resistance. Other than musical traditions, representations of place and people in places in music videos are also relocated into a semiotic package of subversion (meaning to undermine principles and corrupt). These representations are powerful, affecting our understanding of places, reinforcing myths, and providing listeners with a sense of identity (Forman 2002). In song, analysis of settings are 'highly revealing about the world being communicated' (Machin 2010: 92), and 'can be used to understand broader social relations and trends, including identity, ethnicity, attachment to place, cultural economies, social activism, and politics' (Johansson and Bell 2009: 2).

I selected the official video for Grup Kızılırmak's *Çesmi Siyahim* (*My Black Eyed Beauty*) for this chapter. I chose this video first because it clearly articulates protest within the Özgün music genre. Grup Kızılırmak was formed in 1990, producing 13 albums and participating in theatre productions, poetry groups, concerts, and festivals. Its Marxist political orientation is indicated by both performing at Cuban benefits and Marxist conferences and by its music which focuses on Turkey's oppressed, highlighting their plight and offering solidarity

(Grup Kızılırmak 2013). These actions have resulted in its members being arrested and concerts banned or cancelled. Due to the sensitive nature of protest in Turkey, songs even within the mainstream which may be considered subversive are shunned from airplay (Way 2012). However, with the Internet, videos by-pass the broadcasting authorities and are an important part of fans' pop experiences. The video exemplifies this route of distribution. I also chose this video because it was released and popularised when Turkey's Justice and Development Party were in government. Çesmi Siyahim is a cover of a folk song originally performed by Mahsuni Şerif in the 1960s. However, Grup Kızılırmak did not record it until 2002. It appeared on its album *Figan*, while its accompanying video was first uploaded on to the video sharing website İzlesene.com in 2007 and YouTube in 2008.

I offer a short discussion on cultural imperialism and popular music, focusing on Turkey. I then give an historical account of Turkey's politics and mediascape in order to help understand the subversion articulated in the video. This is followed by an analysis of lyrics, visuals, and the music of Çesmi Siyahim. For lyrics and visuals, I examine the representation of place, participants in places, and how visual elements share space to suggest relations between each other. The music itself is also analysed to illustrate how musical elements from both the West and within Turkey are relocated to express protest, the desire for change, and alternatives to dominant discourses offered in Turkish mainstream music and media.

Cultural imperialism and popular music

Economically, Turkey can be described as a country embracing neoliberal capitalism. Its neoliberal economic policies, accelerated over the past decade, include both privatisation and the lifting of restrictions for international economic interests. This applies to most public/state institutions in the areas of communication, transportation, industry, and energy. Symptomatic of this shift is the migration of Turkish pop stars from national labels to the global music giant Polygram (Stokes 2003: 300). Though proponents claim neoliberal style globalisation promotes fair and equal access to world markets, there are fears over the effects this has on local cultures. One fear is 'cultural imperialism' which critics (such as Hall) claim leads to a 'global mass culture', not a two-way balanced flow of culture between 'the West and the rest' (Hall 1992, 1991, Lang cited in Ang 1985: 2). It is attributed to shaping 'social institutions to correspond to, or even promote, the values and structures of the dominating centre of the system [the West]' in non-Western countries (Schiller 1976: 9–10).

Popular music is one area which has experienced globalisation in a form of cultural imperialism more dramatically than other mass communications. Reasons cited are that it is 'less dependent for its comprehension upon language, education, and the acquisition of a sophisticated body of knowledge' than other media, allowing its unhindered global consumption (Negus 1997: 271). This is evident in the spread of 'global' formats from the West such as rock and rap, the global movement of music and instruments, and the 'spectacular inequality in the economic rewards and prestige' of Western pop products outside the West compared with non-Western recordings in the West (Burton 2005: 158; Born and Hesmondhalgh 2003).

Despite these concerns over cultural imperialism, critics including Cunningham and Jacka (1997) question its usefulness as a concept across a variety of media. With reference to pop, it is an oversimplification to make a 'simple correlation' between the global spread of some music and the activities of Western-based multinationals (Stokes 2003: 301). For one thing, these 'majors' do not entirely control music markets due to the role of the independents. Stokes (2003) also claims that reducing the circulation of global genres to the sales of recordings ignores other aspects of music such as live music scenes and listeners' use of music including Internet radio, ipods, and downloading. Finally, the majors have 'real histories of time and place' which shape their way of business including the music they package and sell (Stokes 2003: 301). Other critics of cultural imperialism note that simply mapping the flow of music globally is not enough. Instead, these flows create 'complex patterns of cross-fertilisation and cultural hybridity' (Shepherd 2003: 75). I see this 'cross-fertilisation and cultural hybridity' as the relocating of semiotic resources from centres of power, such as the West, to produce new packages of semiotic meanings. This is evident in Japan where youth 'took rock and did something with it, within their own musical sensibilities and needs for social resistance' (Burton 2005: 158). Punk in non-Western countries is transformed into forms of protest in 'particularised local conditions and circumstances', while rap and hip hop are used to 'reflect and engage with local issues in different cities and regions around the World' (Bennet 2001 in Burton 2005: 157).

Relocating Western musical references in Turkey goes back as far as the later Ottoman era. In 1826, the state formed a new modern army complete with a European-style military band which played a Western repertoire and officially used Western musical notation and instruments. Stokes describes how in the 1970s and 1980s 'rock radicals looked to the Anglo-Saxon World for inspiration' (Stokes 2010: 18–19). In the 1990s, Turkish rap and hip hop 'borrow[ed] energetically from North American models', while the popular genre of Arabesk music is celebrated for its Eastern and Western hybridity. Western sounds and influences are 'localised' by Turkish musicians to express their own concerns. For example, hip hoppers in Istanbul have indigenised rap with local genres of Turkish popular music to 'construct a specifically local identity' (Solomon 2005: 1). As will be discussed below, Özgün music also borrows from the West to subvert the Turkish authorities.

But centres of power can also be local. Turkey's political and Islamist centres of power are being challenged by a more heterogeneous periphery which includes minorities and those opposed to the government's emphasis on Islam and neoliberal economics. Özgün music is one such peripheral voice. Semiotic resources are relocated from the West but also from Turkish mainstream culture to the protesting periphery to enhance messages of subversion. This is seen, for example, in instrumentation and melodies (see below) which have been taken from more government-friendly genres. Özgün music, which emerged with democracy after Turkey's three-year military government following the 1980 military coup, has its roots in Turkish folk music with traditional Turkish instrumentation being a key identifiable element. Though music which may be considered subversive is incorporated into a variety of musical genres, Özgün music is renowned for its politics of protest. This chapter will examine how Grup Kızılırmak relocates Turkish and Western semiotic resources with an emphasis on representations of places and people in places to reveal how they articulate protest against Turkish political centres of power.

Turkish Özgün music and politics

Turkey's Justice and Development Party (AKP) has won three consecutive federal elections since 2002. It is an economically liberal and socially conservative political party founded by members of Islamic and conservative groups. It is seen as an alternative to Kemalism which views Mustafa Kemal Atatürk (the founder of the modern Turkish Republic) as the symbolic leader and hero of Turkey, a champion of secularism, modernisation, and closeness with Europe (Mango 1999). AKP has pursued a socially and religiously conservative agenda, symbolically and legally changing Turkey's secular laws and distancing itself from Western cultural practices. For example, rules on wearing the Islamic headscarf in public institutions which was previously banned have now been relaxed. Also, AKP members do not drink during public functions while they openly talk against alcohol consumption (Hürriyet 2011). AKP emphasises its Islamic roots and future. On its website, AKP claims 'Our party considers religion as one of the most important institutions of humanity'. Religious references are used regularly as in 'May Allah be the beloved and assistant of our nation?' In February 2012, Prime Minister Erdoğan caused controversy by telling reporters that AKP was bringing up a new generation of even more religious AKP politicians (Radikal 2012). By emphasising Turkey's Islamic and non-Western traditions while de-emphasising its European ambitions, many secular Turks are concerned (Mango 1999).

Turkish pop has felt the effects of AKP's time in government. Turkish music production is a multi-million pound industry involving both major global record companies and independents. The majors dominate the market while independent music production is confined to local record labels and distribution. The dominance of the majors has been enhanced by the 1990s liberalisation of broadcasting and the market in general which saw an increase in foreign investment. Stokes notes how Polygram, in collaboration with Turkey's RAK 'signed up nearly every major pop and rock star for a global market by the mid-1990s' while less known groups remained with more 'national' labels (Stokes 2003: 300).

Özgün groups such as Grup Kızılırmak remain locally produced and distributed. Though this may not be a major obstacle for distribution within Turkey, Özgün musicians or musicians deemed to be subversive by the government face many obstacles. There are dozens of examples of censorship (Bülent Ersoy, Ibrahim Tatlises), arrests and exiles (Cem Karaca, Ahmet Kaya, Grup Yorum) resulting from pop's politics. Live performances as well as recordings are subjected to scrutiny and censorship by the government. Authorities can refuse to grant permission for concerts. When granted, concerts may be cancelled at the last minute despite being organised and paid for by the band and its supporters. These actions are usually accompanied by band members being arrested for spreading 'propaganda'. Furthermore, for a recording to be officially released, it must first get a 'bandrol'. This is a sticker issued by the Turkish Ministry of Culture which indicates the product's manufacturer has paid the required tax. However, this branch of the government uses bandrol to censor music, tightly controlling and refusing their issuance for reasons including 'political content, such as song lyrics perceived to advocate violence,[or] political views the government would rather not see expressed' (Solomon 2005: 6).

Another obstacle facing Özgün music is its lack of access to broadcasting. There are over 1,100 radio and 200 television stations, though most are in the hands of state-run TRT and five private media conglomerates. According to the European Journalists Association (EJA), Turkey's mediascape is heavily dominated by large multi-sector groups controlling the media and the distribution of popular music. The EJA observes that '*Kral TV* and *Number One TV* are music channels which broadcast, rank, and promote music-clips and in this way to a certain extent manipulate the music market in Turkey' (Barış 2010).

Relations between these media conglomerates and Turkish politics are very close, characteristic of Hallin and Mancini's (2004) Mediterranean Model. Turkey has a politically oriented press, high political parallelism in journalism, and the state plays a significant role as owner, regulator, and funder of media as it oversees a high degree of ideological diversity and conflict in society. Turkish media conglomerates 'use the media to manipulate other economic or political interests' resulting in a 'notorious interlocking of interests between the media, politicians and the businesses' (Özguneş and Georgios 2000: 414). The resulting mediascape is one of tight political and big business influence and control, even more so than in nations where governments are not as close to the media, like Britain and North America. This situation has been exacerbated by AKP's unprecedented grip on media control where much is owned or controlled by the government and its supporters (Jenkins 2012: 1). It is in this context that Turkish protest music articulates subversion.

Çesmi Siyahim

Çesmi Siyahim's style of protest is not one of direct references seen in Green Day's *American Idiot*, but more suggestive. The visual narrative represents a voyage by the band which travels to a Turkish village where band members witness the working and living conditions of villagers. Then, they travel back to perform in front of a large crowd. Likewise, the lyrics also describe a journey. This time somebody is leaving a place to go to the mountains. While Turkish settings of the road, a village, a concert and mountains, alongside the Turkish language, people and references dominate, Western sources are also utilised to articulate sympathy for the oppressed, commonality with others who protest, legitimacy of protest and a possible alternative to materialism. I examine each of these discourses and how they are manifested in the modes of visuals, lyrics, and music.

Sympathy for the oppressed and specifically the Kurds and the poor can be seen in the visuals and heard in the lyrics and music. Even the name of the band locates Grup Kızılırmak in a part of Turkey with deep Kurdish roots. Naming bands after local names such as Ladysmith Black Mambaso, and Cypress Hill are 'evocative of place, and often history and rurality' (Connell and Gibson 2003: 43). Here, the band is named after the Kızılırmak River in Eastern Turkey. In the past, it was the historic boundary between the Kurds to the East and the Greeks to the West. Today, it is a very poor area populated by many Kurds. Since the band is named as such, Kurdish people, places and their struggles from the past and present are evoked.

The visuals start with a second-person perspective of being on the road. Rural spaces echo mobility in folk and country music, though the trip also draws upon rock'n'roll's idea of moving being about 'freedom and adventure', notable in the shots of the band enjoying their day out (Connell and Gibson 2003: 82). However, both these ideas are usurped by the following series of shots which connote sympathy and support for oppressed people in Turkey, alluding to Turkey's mass internal migration. In this sequence, the band appears in a number of long shots on the move in a poor working village. This is not the adventure evoked in rock'n'roll or a homecoming in Country and Western music. The poor village is a result of Turkey's mass migration from rural to urban centres over the past 80 years which peaked in the 1980s. Migration has 'had a marked effect on development where provision of basic services in rural areas continues to be hindered' resulting in disparities in rates of disease and poverty with 'a much lower standard of living and access to basic services for the smaller rural population' (Unicef 2010). The band walks past and observes villagers who stare at the cameras, look elsewhere, or watch others work. These representations of actions do not affect anyone or anything connoting a lack of power (Fairclough 2003). Others, namely men and boys, engage in poorly paid manual work such as grinding and moulding metal, separating wheat, working in fields and constructing buildings. The settings of decrepit buildings, dark interiors, poor houses and working environments connote poverty, while the dress of both men and women allude to Kurdish and Alevi minority groups. The abundance of close-up shots of villagers grants viewers a point of identification, drawing viewers close and humanising them (Machin 2007: 118–119). These shot choices of minorities in a poor rural setting in the context of a protest song in a minor key (see below) suggest sympathy towards villagers.

Villager shots are montaged together with shots of the band. However, unlike the villagers, the band is always shot as a group within the village, walking or playing. These group shots connote anonymity, less importance or homogeneity (Machin and van Leeuwen 2005: 132). As such, the band suggests the plight of the villagers is important while de-emphasising themselves. Furthermore, how the band shares the visual space in the video suggests relations between the two groups. In all village shots, the band and villagers share common locations of narrow streets, dark spaces where workers perform manual labour, and a small traditional café. However, they rarely share the same shot. By framing the band and villagers in a common location, a loose 'integration' takes place between the two groups which may connote closeness and sympathy. However, by separating them in shots, difference is also articulated (van Leeuwen 2005: 112).

Even when the band and villagers share the same shot, difference is again connoted by the groups not engaging with each other. Band members seem like ghosts, framed in the background of compositions, simply observing. This lack of engagement distinguishes those who are oppressed (workers, minorities, and villagers) from those who fight oppression (the band). In fact, the band is further differentiated from the villagers by having the luxury of mobility, a common theme in country and rock'n'roll, by going on road trips unlike the workers. In one sequence, the group interacts with villagers symbolically and literally. Here, the band drinks Turkish coffee in a traditional coffee house. The colourful patterned cushions, off-white walls and decor make it clear the place is a traditional workers' 'kahvehane', not an upmarket café. A waiter interacts with the group by exchanging glances and pouring coffee into their cups. Interestingly, the coffee 'cevze' (vessel which the coffee is made) looks like the cevze the group observed being made previously. This sequence links the worker who makes the cevze with the waiter with the band, creating a bond of understanding about the plight of workers and minorities in rural Turkey.

Sympathy for the oppressed can also be heard in the lyrics. 'My wealth (capital) is my pain/concern my friend, My wealth is my curse/frustration' is repeated twice. Collocating wealth and capital with the nouns 'pain', 'curse', 'concern' and 'frustration' four times expresses a negative attitude to wealth. Though circumstances are unclear, a common trait of pop (Machin 2010: 92), one can speculate that like the oppressed, wealth is not something they have the luxury to enjoy, but its scarseness is a 'curse', 'pain', 'concern' and 'frustration'. Here, the narrator aligns herself with those who are oppressed, in this case financially, perhaps experiencing poverty similar to the villagers in the visuals.

During these verses, minor and major chords along with the melodies played by guitars and the bağlama further suggest sympathy towards minorities and workers. The song is in G minor; minor keys in Turkish and Western music used to symbolise sadness and remorse. With lyrics and visuals articulating sympathy for Alevi and Kurdish villagers and workers, the choice of the minor key enhances sympathy and illuminates the harshness of their plight. For example, the musical sequence in the introduction is Gm, Fmaj, Am, Cmaj to Dm. Two characteristics of this sequence suggest sympathy. First, melodies with descending pitch 'can relax and soothe listeners, make them turn inwards, and focus on their thoughts and feelings' (van Leeuwen 1999: 103). Here, pitch descends, signalling to listeners that this song is one for thought. Accompanying visuals which show a poor Alevi village focus such thoughts. Second, melodies with a narrow pitch range, such as in the introduction, 'constrains the expression of strong feelings' whether due to modesty, tiredness or fear, to mention just a few possibilities (van Leeuwen 1999: 106). Here, the narrow pitch range symbolises the style of protest seen throughout the song. It is not one with flowing emotions, but one of constrained sorrow and the poetic articulation of thoughtful alternatives to Turkey's present condition.

The band shows their support for not only the oppressed, but also those who protest. In the visuals, this is seen in the kahvehane sequence where the band shares the same location with a picture of Yilmaz Güney, a film-maker, actor, communist and protester who was imprisoned in 1961 for publishing a communist novel. From this time until his death in 1984, Güney struggled with the Turkish authorities resulting in imprisonment, escape, and then fleeing to France before having his citizenship revoked. In the visuals, while the band drinks coffee, a picture of Güney hangs in the centre of the background. It is salient, being in the centre of a mostly plain background. The two visual elements (band and Güney) are linked (van Leeuwen 2005) by the band sharing a space with the picture. This shot is followed by a full-screen close-up of the same picture of Güney. This close-up aids the viewer in identifying Güney and emphasises his importance to the band (Machin 2007).

Lyrics also suggest support for those who protest in the lines 'You laughed at the poor condition of Mahzuni my friend, Even if he perished in the hands of strangers'. Aşık Mahzuni Şerif was a Turkish protest folk singer, poet and author. 'Aşık' indicates his position as a respected

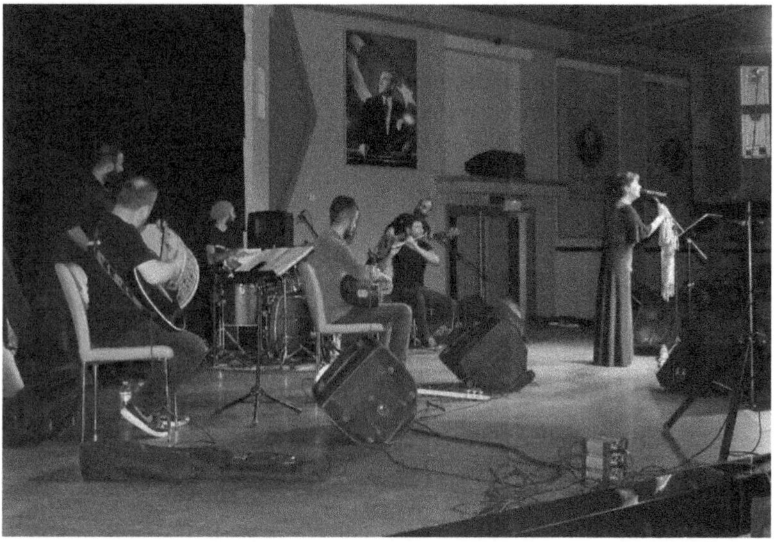

Figure 1.1 West and Turkish relocations in concert

musician and his Alevi religious affiliations. Şerif suffered intimidation by having his house burnt down, being investigated and accused by Turkey's State Security Court for protesting against the government including declaring 'Hem Kızılbaş, hem Aleviyim' (I am both Kızılbaş and Alevi), before fleeing to Germany in 2001 where he died. Until 2000, both Kızılbaş and Alevi sects of Shi'a Islam were not allowed to openly practise their religion. In the lyrics, 'you' antagonise Mahzuni by laughing at his poor condition, just as the Turkish authorities antagonised him. Though this representation lacks detail, 'you' are represented negatively and opposed to the group of 'us' who are oppressed, protest and subvert. These representations suggest support for protesters and subversion towards Turkish authorities.

The band is also represented as protesters themselves. Again, this mostly works on a symbolic level. Clothes, hair, and instruments are important semiotic resources for bands (Machin 2010). These resources are used to connote associations with various genres, each with their own set of politics. So, for example, bands may choose a rock aesthetic to articulate being anti-establishment and rebellious (Frith 1981; Machin 2010; Street 1988). Band members do not dress traditionally like the villagers, but relocate symbols associated with Western rock into the village scene. Some of the men have long hair, sideburns, wear jeans, denim and leather jackets, while black and other dark colours are dominant. These same style choices can be seen in rock imagery from the Rolling Stones' 'Black and Blue' album cover to most images of Motörhead, the Stranglers, and Muse. References to Western rock go beyond clothes and hair. Electric guitars and a full Western rock drum kit are seen both in the village setting and in the concert. In both settings and in some shots, they are salient, dominating the foregrounds and connoting their importance. Here, Western rock symbols are relocated into a protest song and used as symbols of subversion to those in Turkey who promote a more Eastern outlook, including the government.

Protest and subversion is also alluded to in the lyrics. In the first lines of the first two verses, the narrator sings of journeys involving 'mountains' with 'vineyards'. References to the outdoors draw upon folk aesthetics and authenticity, favouring rural areas which produce an 'invented geography ... of a bygone natural environment' (Connell and Gibson 2003: 39). Though 'vineyards' and 'mountains' help authenticate the song within the folk genre, it also expresses protest in the context of Turkish Özgün music. The use of place in reggae draws imaginary maps which become symbols of a movement from suffering to redemption (Daynes 2009). Here imaginary maps are drawn to symbolise subversion. In verse one, the narrator talks of mountains in terms of separation in 'My departure means mountains between us'. This metaphor suggests great distance between the narrator and the friend who betrays her later in the lyrics. In verse two, the narrator invites the listener to join her journey in 'Come on, my friend let's go for a walk over the high mountains'. In Turkish literature and history, people who experience problems with authorities escape 'to the mountains'. Here, they become fugitives or bandits, taking up guns, some stealing, and some fighting oppressive authority figures. In some cases, these characters have been romanticised, represented as heroes similar to *Robin Hood* (see *Yalnız Efe* by Ömer Seyfettin). There are also present-day connotations associated with mountains. Kurdish citizens are represented in Turkish nationalist discourses as 'mountain Turks' (Mumcu 1995: 155–156; Tusalp 1988). And it is in the mountains of south-east Turkey where Kurdish Nationalist fighters battle with the Turkish military. The narrator's 'going/walking' to the 'mountains', alongside visuals of Alevi and Kurdish suggests villagers' opposition to authorities. This opposition can be understood either historically as in characters like *Yalnız Efe* or in terms of modern opposition (political and armed) by Kurds in mountainous rural Turkey.

Musical key choices during these verses represent resistance positively. The song is in a minor key, symbolising sympathy and support for the plight of the oppressed. However, between each line of sung lyrics there is a short musical solo in Cmaj which plays a DFEBCC#D sequence. This sequence articulates more positive themes than the rest of the verse. First, the major key connotes something positive, unlike a minor key. Second, part of this sequence (BCC#D) draws on jazz chromatics, a reference to Western music rarely found in Turkish music. With accompanying lyrics which refer to mountains and the possibility of resistance, maybe even armed resistance, the video works multi-modally to articulate the positive possibility of resistance for the victims of oppression.

The concert sequences also legitimise protest while authenticating the band through rock aesthetics. Rock's authenticity is closely related to live performance and 'It is live rock which binds audience to musician' (Frith 1981: 80). Though Grup Kızılırmak is not a rock act, a similar criteria of being authentic through live performance is used in Özgün music. Furthermore, choosing these images suggests a degree of respect in terms of the band being functionally identified as professional musicians (van Leeuwen 1996: 59). This legitimacy is furthered by shots of a large enthusiastic crowd which reacts to the band and their political messages by clapping, cheering, and dancing. It also signifies the band's power, being able to produce such reactions. Though we cannot see where the concert is held, the size of the crowd and their Western dress suggest this is a city, not a poor village. Urban dwelling is an important part of authenticating bands within the rock genre (Connell and Gibson 2003: 37). Perspective and camera angles are used to connote the band's legitimacy and power. Band members are shot both individually in close perspective and as a group. The close ups connote intimacy and support for the band, while group shots connote team spirit (Machin and van Leeuwen, 2005). In most shots, the camera is tilted up connoting power and legitimacy. Alternatively, audience shots are taken from above connoting less power (Machin 2007: 114). Also, there are no individual shots, the audience is always represented as a large homogenous group unified in their approval of Grup Kızılırmak (Machin 2007: 118). Like the villagers, the audience has limited power (in this case, responding to the band), shares the same space as the band, and sometimes appears in the same shots. However, unlike the villagers, the audience is represented positively as happy, enthusiastic and energetic, not oppressed. In this visual hierarchy of audience and band, the audience may symbolise hope of subverting oppression through protest which is articulated and led by legitimate protesters like Grup Kızılırmak.

Grup Kızılırmak offer clues as to its alternative to neoliberal economics in a visual sequence of flying seagulls. Fans of Grup Kızılırmak told the researcher the importance they place on the seagull and specifically the main character from the fable in novella form of *Jonathan Livingston Seagull – a story* written by American writer Richard Bach (1970). The book's main character, a seagull, was frustrated with meaningless materialism in the form of daily fights for food, so became an outcast to pursue the art of flying. One fan quoted the line 'You've got to understand that a seagull is an idea of unlimited freedom' reminding the researcher that the story is a metaphor for freedom from materialism. Shots of flying seagulls in bright open skies are in stark contrast to the oppressive closeness and lack of light in many of the village scenes. The shots draw upon folk and country aesthetics of an 'invented geography ... of pristine remnant[s] of a bygone natural environment' (Connell and Gibson 2003: 39). However, again these are used to symbolise subversion, in this case the freedom fans can strive for through changing society towards a less materialistic one. Instead of concerning themselves with petty materialism, the oppressed can rise above their position through protest. Here, a Western literary reference is relocated into a Turkish protest song and used as a symbol of subversion to governmental economic policies.

Alternatives to Turkey's social conservatism are also suggested through the integration of musical elements. At its subversive roots is the notion of integrating the West with the East. This idea draws upon Turkey's Kemalist political and social modernisation project which champions secularism, modernisation, and closeness with Europe (Mango 1999). Presently in Turkey, much of AKP's policies and ideas differ from Kemalism, emphasising social conservatism and Islam, as already mentioned (AKP 2013). Choosing to include an abundance of Western as well as Turkish musical references may be read as a musical protest against AKP's more traditional 'Eastern' and Islamic outlook. This can be seen in both instrumentation and melody. Though the original song by Mahzuni Şerif is a traditional folk song, Grup Kızılırmak use both Turkish and Western instruments. For example, in the introduction and in musical interludes throughout the song the traditional bağlama plays alongside the guitar. One sound does not dominate the other. Both are prominent in the hierarchy of sound, vying for attention and suggesting equal importance. The bağlama may reference the song's traditional past while the electric guitar refers to more modern and Western rock sounds. The use of rock in this context is subversive whereas in some contexts in the West it may be perceived as anything but, highlighting the importance of context. The foregrounding of both these instruments can be read as a metaphor for Turkey in the East 'looking West', subverting AKP's inward-looking emphasis on Turkishness and Islam.

The choice of using Western tonal harmony is another relocation from the West which can be read as a symbol of subverting AKP. In Turkish music, both traditional art and folk music are based on modality, unlike Western music which is based on tonality. In modal music, scale alone is insufficient and melodic progression is necessary in order to define a certain mode. Furthermore, while certain pitches of the scale have tonal functions in Western music, in modal music pitches of a mode do not have such tonal functions (Ayangil 2008). Using Western tonality is a widespread syncretic musical discourse seen in much Turkish political music (Gedik 2010). By Grup Kızılırmak relocating Western tonal harmony alongside the song's many Turkish characteristics, East and West integration is again suggested.

The mixing of East and West references also calls into question the idea of a homogeneous Turkey. This is seen in instrumental choices between the second and third verses. Here, a cello which is an instrument firmly placed in Western music tradition, is foregrounded in the hierarchy of sound. It plays a slow progression of notes similar to the song's Turkish folk roots. This 'mixed' cello sound foregrounds guitar and bağlama and a complex and fast paced rhythm played by the bass and drums reminiscent of jazz. This more complex instrumentation and musical arrangement may refer not only to an East-West integration, but also to the complex cultural background of Turkey. Like most nations, the Turkish government tries hard to homogenise everyone as 'Islamic' and 'Turks' (Billig 1995). Difference within Turkey is de-emphasised, ignoring its many ethnic groups including Kurds, Armenians, Lars, Benetines and Alevi. This is evident in General Abdullah Alpdoğan assertion in 1936 that 'the people of Eastern Turkey are mountain Turks and Kurdish is the mountain accent of Ottoman Turkish' (Mumcu 1995: 155–156). This same discourse is repeated throughout Turkey's history (Tusalp 1988). Even today, the AKP tries to homogenise its citizens as Turkish, conservative and Muslim (Hürriyet 2011). By exploiting diverse musical references through instrumentation and arrangement, Grup Kızılırmak connotes diversity in Turkey which subverts AKP's homogenising efforts.

Conclusion

Protest music has a long history in Turkey, whether Leftist politics in the 1970s, Anadolu rock or cultural politics in Arabesk and Kurdish language music. Özgün music, with its roots in Turkish folk music continues this tradition. Grup Kızılırmak's popularity and long history has not hampered its radical politics as it continues to articulate support for those oppressed by the Turkish state such as Kurds, Alevi, the poor and those who stand up to the authorities. This chapter has examined how this is done by both relocating Turkish and Western resources including representations of place to a space of resistance.

Instrumentation, melodies, genres and literary references are taken from their Western or Turkish origins and relocated into Turkish protest across the three modes of visuals, lyrics, and music. This is not an example of cultural imperialism which leads to a global mass culture. On the contrary, the case of Grup Kızılırmak demonstrates that pop can harness both local and Western references to articulate discourses which support those who are oppressed and protest against their oppressors. At the same time, Grup Kızılırmak harnesses the power of representations of place and people in places in lyrics and visuals to subvert Turkish centres of power. Here, the band draws upon folk's constructed, imaginary, and ideal rural settings and rock's longing for being on the road and in the city. Grup Kızılırmak, however, uses these not only to authenticate itself as a Özgün protest band but also to articulate subversion to Turkish authorities.

Such protesting takes on critical importance in Turkey where protest is discouraged and even punished, notable in the large number of students, activists, journalists and musicians languishing in Turkish jails. This lack of freedom has been exacerbated by the current government's unprecedented grip on media power where most media companies are owned by the government or its supporters. However, by articulating protest through relocating representations of places and other Turkish and Western resources, Grup Kızılırmak are able to articulate very national injustices and allude to possible solutions.

Works cited

AKP (2013). *AK Parti: Parti Programı*, http://www.akparti.org.tr/site/akparti/parti-programi, accessed 25 April 2013.

Ang, Ien (1985). *Watching Dallas: Soap Opera And The Melodramatic Imagination* (London: Methuen).

Ayangil, Ruhi (2008). 'Western notation in Turkish music', *Journal of the Royal Asiatic Society* (Cambridge: Cambridge University Press).
Bach, Richard (1970). *Jonathan Livingston Seagull: A story* (London: Harper Collins).
Barış, Ruken (2010). *Media Landscape: Turkey, European Journalism Centre*, http://www.ejc.net/media_landscape/article/turkey/, accessed 19 December 2011.
Billig, Michael (1995). *Banal Nationalism* (London: Sage).
Born, Georgina and Hesmondhalgh, David (2003). *Western Music and its Others* (London: University of California Press).
Burton, Graeme (2005). *Media and Society* (Maidenhead: Open University Press).
Connell, John and Gibson, Chris (2003). *Sound Tracks: Popular Music, Identity, and Place* (London: Routledge).
Cunningham, Stuart and Jacka, Elizabeth (1997). 'Neighbourly relations? Cross-cultural reception analysis and Australian soaps in Britain', in Annabelle Shreberny-Mohammadi, Dwayne Winseck, Jim McKenna and Oliver Boyd-Barrett (eds), *Media in Global Context* (London: Arnold), pp. 299–322.
Daynes, Sarah (2009). 'A lesson in Geography, on the Riddim: The symbolic Topograpjy of reggae music', in Ola Johansson and Thomas L. Bell (eds), *Sound, Society And The Geography Of Popular Music* (Ashgate: Farnham), pp. 91–106.
Fairclough, Norman (2003). *Analysing Discourse: Textual Analysis for Social Research* (London: Routledge).
Forman, Murray (2002). *The Hood Comes First: Race, Space, and Place in Rap And Hip-Hop* (Middletown: Wesleyan University Press).
Frith, Simon (1981). *Sound Effects* (New York: Pantheon).
Gedik, Ali (2010). 'Turkiye'deki Politik Muziğin Değişimi uzerine Bir Tartişma Cercevesi: Ey Ozgurluk! Gelenek', *Aylik Marksist Dergi*, 107, pp. 61–80.
Grup Kızılırmak (2013). *Grup Kızılırmak Biyografi*, http://www.ilkayakkaya.com.tr/?page_id=7, accessed 21 March 2013.
Hall, Stuart (1991). 'Old and new identities, old and new ethnicities', in Anthony King (ed.), *Culture, Globalisation and the World System* (London: Macmillan).
Hall, Stuart (1992). 'The west and the rest: Discourse and power', in Stuart Hall and Bram Gieben (eds), *Formations of Modernity* (Cambridge: Polity Press), pp. 275–320.
Hallin, David and Mancini, Paulo (2004). *Comparing Media Systems: Three Models of Media and Politics* (Cambridge: Cambridge University Press).
Hürriyet (2011) *Babalar Burada,* http://www.dha.com.tr/tusiad-basbakan-recep-tayyip-erdogan-alkol-tartismasi-yasam-tarzi-otomobil-son-dakika-haberi_137188.html, accessed 29 March 2011.
Jenkins, Gareth (2012). 'A house divided against itself: The deteriorating state of media freedom in Turkey', *Central Asia-Caucasus Institute Silk Road Studies Program,* http://www.silkroadstudies.org/new/inside/turkey/2012/120206A.html, accessed 02 February 2012.
Johansson, Ola and Bell, Thomas L. (2009) 'Introduction', in Ola Johansson and Thomas L. Bell (eds), *Sound, Society and the Geography Of Popular Music* (Ashgate: Farnham), pp. 1–6.
Machin, David (2007). *Introduction to Multimodal Analysis* (London: Hodder Education).
Machin, David (2010). *Analysing Popular Music* (London: Sage).

Machin, David and van Leeuwen, Theo (2005). 'Computer games as political discourse: The case of Black Hawk Down', *Journal of Language and Politics* 4(1), pp.119–141.

Mango, Andrew (1999). *Atatürk: The Biography of the Founder of Modern Turkey* (London: John Murray).

Mumcu, Uğur (1995). *Kurt Dosyasi* (Ankara: Tekin Yayinlari).

Negus, Keith (1997). 'Global Harmonies and Local discords: Transnational policies and practices in the European recording industry', in Annabelle Shreberny-Mohammadi, Dwayne Winseck, Jim McKenna and Oliver Boyd-Barrett (eds), *Media in Global Context* (London: Arnold), pp. 270–283.

Özguneş, Neslihan and Terzis, Georgios (2000). 'Constraints and remedies for journalists reporting national conflict: The case of Greece and Turkey', *Journalism Studies* 1(3), pp. 405–426.

Radikal (2012). Erdoğan *'dindar nesil'i savundu*, http://www.radikal.com.tr/Radikal.aspx?aType=RadikalDetayV3&ArticleID=1077899&CategoryID=78, accessed 21 February 2012.

Schiller, Herbert (1976). *Communication and Cultural Domination* (New York: International Arts and Sciences Press).

Shepherd, John (2003). 'Music and Social categories', in Martin Clayton, Trevor Herbert and Richard Middleton (eds), *The Cultural Study of Music: A Critical Introduction* (London: Routledge), pp. 69–80.

Solomon, Thomas (2005). 'Living underground is tough': Authenticity and locality in the hip-hop community in Istanbul, Turkey', *Popular Music* 24(1), pp. 1–20.

Street, John (1988). *Rebel Rock: The Politics of Popular Music* (Oxford: Basil Blackwood).

Stokes, Martin (2003). 'Globalisation and the politics of World music', in Martin Clayton, Trevor Herbert and Richard Middleton (eds), *The Cultural Study of Music: A Critical Introduction* (London: Routledge), pp. 297–308.

Stokes, Martin (2010). *The Republic of Love: Cultural Intimacy in Turkish Popular Music* (London: University of Chicago Press).

Tusalp, Erbil (1988). *Eylül Imparatorluğu, Doğusu Ve Yükselisi* (Ankara: Bilgi).

Unicef (2010) *Unicef in Turkey: Country Profile*, http://www.unicef.org/turkey/ut/ut2_2010.html, accessed 27 May 2014.

van Leeuwen, Theo (1996). 'The representation of social actors', in Carmen Caldas-Coulthard and Malcolm Coulthard (eds), *Texts and Practices – Readings in Critical Discourse Analysis* (London: Routledge), pp. 32–70.

van Leeuwen, Theo (1999). *Speech, Music, Sound* (London: Macmillan Press).

van Leeuwen, Theo (2005). *Introducing Social Semiotics* (London: Routledge).

Way, Lyndon (2012). 'Turkish popular music videos as a multi-modal site of resistance', *Multimodal Communication* 1(3), pp. 251–275.

2
Sampling the Sense of Place in Baile Funk Music

Sandra D'Angelo

Baile funk[1] is the first manifestation of Brazilian electronic dance music (Palombini 2010: 103; Sá 2007) to emerge from the slums area of Rio de Janeiro in the late 1980s.[2] In terms of its cultural impact, this genre offers a strong sense of place: a pictorial place of art and dreams, a meaningful place of action and, finally, a utopic place for thousands of socially excluded youths living in harsh conditions in Brazilian shanty towns. The utopian dimension in baile funk music is articulated through its multilayered musical nuances including *proibidão,* regarded as a way to celebrate prohibited powers or *romântico,* which exalts a utopic world of romanticism and more recently, the utopian is represented in *funk ostentação* (funk ostentation). I approach the phenomenon of baile funk as a manifestation of a postmodern 'loss of the real' in the way its narrative can be read as a way to disconnect with the *real* condition of the *favela*. Through the process of transnational relocation of music, the genre can also be seen as a postcolonial conjointment in the way, baile funk 'writes back' (to borrow Ashcroft's concept) to Europe and the USA to affirm its own sense of place, and to claim 'the right [as a cultural product] to represent [itself] by juxtaposing colonial inscriptions' (Fiedler 2007: 275). In Brazil, this genre is usually called funk carioca, favela funk or just funk, but throughout the chapter I refer to it as *baile funk* as it is called among musicians outside Brazil playing the genre. The genre reinterprets the sense of place that has been transnationally relocated by baile funk through the process of 'writing back'. Baile funk usually exhibits a fixed tempo of around 126bpm to 134bpm (D'Angelo 2014), and has at its core the *sample,* a portion of sound extracted from an existing song recorded from the environment or synthetically generated by a sampler.

The study of baile funk is a relatively new academic phenomenon and has been approached mainly from a sociological perspective, although concepts such as music and place are part of a wider body of academic literature. However, baile funk music has never been approached in terms of its construction of a sense of place. At the core of my research, insights emerge from interviews collected between 2010 and 2013, and through participant observation I conducted throughout a year while living in Rio de Janeiro in the favela of Pereira da Silva (*Pereirão*) and in the *Laranjeiras* in 2010. I also investigate discussions on Brazilian and European Internet forums for baile funk followers from 2008 to 2014. These include Internet weblogs such as *Faveladowordpress* or *Riobaile funk.net* which have become common networking sources among baile funk DJs and enthusiasts. A total of 43 face-to-face interviews were conducted with baile funk DJs, producers, MCs, and followers of baile funk and other Brazilian pop genres in Rio de Janerio (five MCs, 10 DJ producers, five baile funk followers and 10 non-followers), São Paulo (three DJ producers, seven non- followers) London (seven DJ producers and promoters), Paris (one promoter, two DJ producers) Berlin (two DJs and promoters), Italy (three DJ and promoters). The interviews lasted between 30 and 60 minutes. The majority of interviewees in Brazil were of brown or black lower-class origin and their ages ranged from 19 to 59 years. In Europe, all interviewees were white, middle-class and aged between 30 and 50 years. They were selected on the basis of being well-established DJs, or music enthusiasts with a variety of opinions on both baile funk music and the sense of place created and delivered through this genre.

The utopian and topophobic dimensions of Baile funk

Baile funk surfaced in Rio de Janeiro in the late 1980s owing to easier access to international music records and the importation of more affordable music technology (Palombini 2010: 103). Most of baile funk's production occurred in small-scale studios in favelas from where the music was first distributed through self-produced cassettes and later via CDs in markets throughout Rio and all over Brazil. The first album of the genre, *Funk Brasil* (PolyGram Brazil), was produced by DJ Marlboro in 1989. Today, baile funk has become representative of favela life, a point put forward by Livio Sansone:

> [...] funk is an increasingly pervasive aspect of Brazilian society. As any other researcher in the field working in a lower-class community

> in Rio in the mid-1990s, I was immediately confronted with the omnipresence of funk music in the alleyways that form the main part of this self-built community. Walking through those alleyways, the predominant sound is the bass of the drum machine, used in funk tracks, through radios or cassette players. The sound of samba is decidedly less present. (Sansone 2003: 119)

and Robert Neuwirth who writes of how:

> All across the favelas, few people listened to the music that outsiders think of as Brazilian. Everyone knows the samba, bossa nova, and Musica Popular Brasileira (MPB) hits. They're the soundtrack of the telenovelas [...]. But the mass of favela dwellers have embraced hard core rap and funk [...] as their emblematic sound. (Neuwirth 2005: 39)

In terms of the origins of the concept of baile funk as dance music, this developed in Rio de Janeiro during the 1970s from *baile black*[3] (black dances) or *black Rio dances*. In 1976, the journalist Lena Frias was the first to name and report the baile black phenomenon in an important national journal. These dances, organised by *equipes de som* (sound teams) in the shanty towns of Rio de Janeiro, 'used to attract each weekend from five hundred thousand to one and half million' youths (Palombini 2010: 99). The typical music played in Bailes Black[4] was soul and funk music produced and diffused during the 1960s and 1970s by artists such as James Brown and Parliament-Funkadelic. From the late 1980s, the music played at baile funk dances in the suburbs and favelas radically changed and shifted from North American funk to North American electro-funk.

A negative perception of the favela and favela culture is generated throughout the section of Brazilian media, which describes these areas as deprived of essential public services.[5] According to IBGE (the Brazilian Institute of Geography and Statistics), favelas are 'subnormal agglomerations of shacks and houses'. The favela could be conceived of as a city within a city, a place where the community has had to survive almost entirely autonomously because it has suffered and still suffers from stigmatisation and exclusion from the rest of Brazilian society.

It is not surprising that some Brazilian musicians from other musical genres perceive baile funk music in negative terms. For instance bossa nova musicians, whose origins are in the Brazilian white middle or upper social classes, are also hip hoppers, who share the same social and economic conditions but refer to a different philosophical understanding and purpose of music. Other detractors include Miami bass musicians from Rio de Janeiro, who consider baile funk as a bad copy of Miami bass and *sambistas,* from whom *funkeiros* (baile funk musicians and enthusiasts) inherited the same physical spaces and social conditions but had an antithetic *sense of* music production and creative impetus. To explain the controversial relationship between the place *favela* and part of Brazilian society, it is worth highlighting substantial socio-political and historical differences occurring between samba and baile funk. Although they share the same socio-economic conditions and urban environment, they are perceived as diametrically opposed to one another. To do so, I shall-offer an overview of the historical context.

Samba and the idealised view of Brazil

The ideology of the Brazilian dictator Getulio Vargas *Estado Novo* (New State) was imposed by a revolution that destroyed the constitutional government and established totalitarianism during the so-called *era Vargas* or *periodo Getulista* (1930–1945). A specific sense of 'Brazilianness' (Perrone and Dunn 2001), was delivered to Brazil's main economic partner the USA, during the 'Good Neighbour' policy (in the 1930s), through the on-screen manifestation of Brazilian movie stars such as Carmen Miranda (Ovalle 2011: 61) and popular music such as samba. The dictator contaminated the popular samba scene by censoring and promoting only the patriotic verses of the subgenre *samba-exaltação* (patriotic samba) (Yudice 1994). According to Reily, 'The onset of the Vargas regime coincided with the emergence of samba, a hybrid popular style that provided the Estado Novo with a ready-made musical form well suited to the nationalist agenda' (Reily 2000: 10).

An example of patriotic samba is *Aquarela do Brasil,* written in 1939 by Ary Barroso, a composer from a white lower middle-class background (Shaw 1998). Barroso a fervent supporter of Vargas like the singers, 'Orlando Silva, Francisco Alves and others, [was] co-opted by the state-owned National Radio to sing the glories of Brazil' (Reily 2000: 10). During his regime, Vargas invented and exported a specific sense of 'Brazilianness' (Perrone and Dunn 2001), utilising the concept of *mestiço,* which emerged during the 1930s in Gilberto Freyre's *The Masters and the Slaves* (1966) (Reily 2000: 10). The hybrid mestiço, emblem of the Brazilian as a 'cordial' man was 'heralded as the symbol of national unity, embodying the synthesis of the country's disparate social categories' (ibid.). The 'friendly', 'happy', 'cordial' and 'industrious' mestiço was 'soon to become a common-sense category with the population at large' (Reily 2000: 10). For all these reasons, samba became a fundamental part of the official view of 'Brazilianness'.

By contrast, baile funk 'questions the fantasy of access to social space for the underprivileged and the convivial image that those in power wish to propagate of Brazil' (Yudice 1994: 197). Discussing class aspects of baile funk, George Yudice points out in 'The Funkification of Rio' that working-class youths identify themselves more with the local favela instead of the nationalistic connections of musical genres such as samba. Baile funk culture is considered by the Brazilian mainstream as an antisocial cultural phenomenon with a dangerously provocative attitude probably because it destroys the mystique of a 'happy' relationship between the upper and the lower classes (Yudice 1994). Throughout its music aesthetic and approach to the representation of day-to-day reality in the favela, the genre expresses a rejection of the fake promise of a better life through the acceptance of state sanctioned social participation and civil contribution (Yudice 1994) (Figure 2.1).

Going back to the analysis of the favela life conditions and the significance of baile funk for people, baile funk, in line with Paul Sneed (2003, 2008), can be read as a mode of expression for those reacting

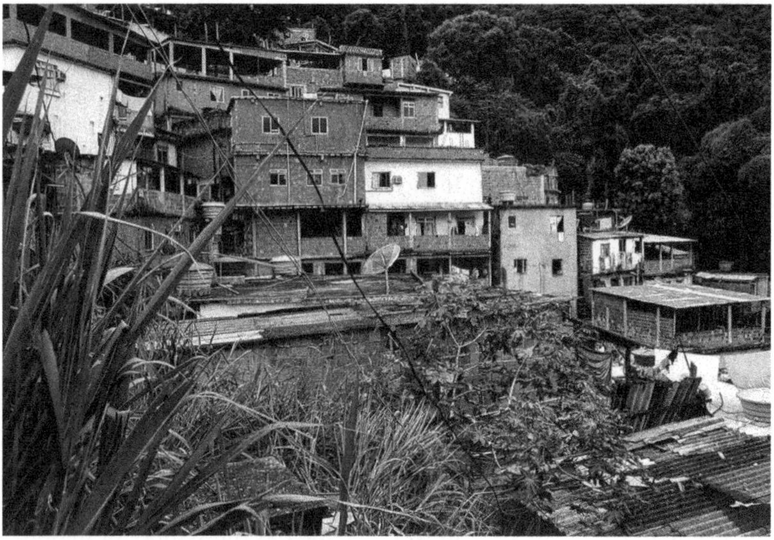

Figure 2.1 Favela of Pereira da Silva (*Pereirão*) Rio de Janeiro
Source: Photo Sandra D'Angelo. Rio de Janeiro, 2010.

to harsh living conditions. In favelas, homes are made with the most basic materials. People do not have access to decent schooling and have limited or non-existent access to public services such as sanitation, electricity, and public transport (Dennison 2005: 352). Within the favelas, criminals have produced a culture of silence through violent intimidation and high mortality is commonplace; 'Of the 3,000 deaths that occur every year in Rio de Janeiro, most take place in the favelas' (ibid.). Furthermore, police regularly invade favelas in search of drug dealers who disappear in labyrinthine streets (ibid.). Thus, in contrast to the claustrophobic and violent realities of life (physical and psychological), baile funk, read as a collective activity, offers freedom through the construction of an alternative *space*. This view is supported by Sneed who suggests that baile funk is a temporary collective space where utopic impulses are revealed through community participation which subverts the mainstream aesthetic, values and modes of social interaction. People, in the baile funk party experience feelings of power, courage, and hope. 'These feelings are accompanied by a sense of inclusion in the larger Brazilian society, not necessarily the Brazil of the middle and upper classes, but a Brazil of everyday people, an intensely Brazilian space of hyper-brasilidade, in all its racial openness, sensual playfulness, musicality, gregariousness, and merriment' (Sneed 2008: 77).

The semiotic codes of Utopia

Dyer in his essay 'Entertainment and Utopia' identifies music as the most 'felt' utopian practice (Dyer 1999: 373). Regarding utopianism, Dyer further points out that 'entertainment does not, [however], present models of utopian worlds, as in the classic utopias of Sir Thomas More, William Morris, *et al.* Rather the utopianism is contained in the feelings it [the entertainment] embodies' (Dyer 1999: 373). As regards the concept of utopia and entertainment, although there are similarities between Sneed's and Richard Dyer's theories, a fundamental difference is evident in the explication of the dynamics between production and consumption. Dyer discusses utopia and professional entertainment within the socio-political context of North American musicals. For Dyer, the audience consumes media products to escape social tension, inadequacy and absence, and so professional entertainment provides a utopian dimension and an escapist solution for day-to-day existence (Dyer 1999: 20). However for Sneed (2003, 2008), especially in baile funk events, the utopic dimension can be reached only through the active involvement of the audience during the performance whereas for

Dyer, professional entertainment does not necessarily involve this level of active engagement and audience participation.

In the light of the above discussion, I would argue that baile funk reveals its utopic nature through its artistic compositional features on the strength of affective and semiotic codes, and the use of both representational and non-representational signs. On account of that, I understand the creation of baile funk's utopic dimension to be a specific sampling aesthetic. In this aesthetic, both representational signs, particularly words and lyrics, converge with non-representational signs (particularly rhythms, melodies and sonic fragments) (Dyer 1999) which constitute a music producer's sonic palette on any given track. Sampling, as the act of extracting, storing and processing fragments of sound from diverse sources and the reworking of them through sound effects such as reverb, are aesthetic acts that rely on a semantic reinterpretation of the meaning and sense of place from where the sound originated.

For all these reasons, I argue that baile funk's utopic dimension is created by producers through sampling practice and through the individual 'grain' of samples, which can include both representational signs such as lyrics and spoken words, and non-representational signs such as sonic fragments signifying emotional connotations associated with local or foreign places. The grain of a sample reveals the qualities of a fragment of sound, which in baile funk reflects the producer's attempt to create a nuanced pictorial sonic alphabet capable of eliciting an emotional response from the audience. One of the most popular sounds in baile funk music is the sample of a revolver gunshot often used by DJs, in terms of music production, to empathise and accentuate the rhythmic patterns of a song. Through a comparative content analysis of several fragments of gunshots sounds, I conclude that the sample was extracted from a song *'O Homem Mau'* (the Bad Man), that is part of the traditional music *sertaneja* from the Brazilian duo Leo Canhoto and Robertinho, very famous during the 1960s. I would argue that this sample manifests the postmodern aspect of baile funk music in the way it quotes the past as well as the present.

The sample's postmodern qualities are also revealed in the way its intertextuality reworks the meaning of traumatic events in the favela soundscape. It is not rare for innocent residents in suburbs or favelas to be caught in the crossfire between police invading their homes in search of criminals. In the light of such incidents, baile funk musicians intentionally use the gunshot sample in order to subvert its semantic order (Sneed 2008) and dissolve it in a sort of cathartic reduction of the tension and the anger. The same modality appears among ordinary residents of the favelas that through a specific gesture, the gun and its sound are simulated altering its semantic connotation. '[In the baile funk dance] the little boy makes a gun with his thumb and fingers and continues smiling as he gestures along with the digital gunshots of the MC's song – bang, bang, bang!' (Sneed 2008: 59)

The use of the gunshot sample evokes a traumatic spatial impression, that DJs juxtapose on a positive sense of puissance and humour in order to dominate, address, and decrease chronic social tension in the audience, which is unfortunately part of everyday favela reality (Figure 2.2).

In the broadest terms, a sense of place is built through emotional/affective codes and semiotic signs; it is constructed via specific associations to places, mediated by the complex nuance of emotions that I summarise here as repulsion or affection. Place researchers such as Yi-Fu Tuan (1974, 1977, 1980) and Relph (1976) have long theorised the emotional connection between people and places. Following Relph, who claims that a sense of place is determined by the activities that happen there, we can say that music is one of the rituals/activities most important to groups of people, and so the creation of music contributes to constructing and representing a sense of place and a 'place identity'.

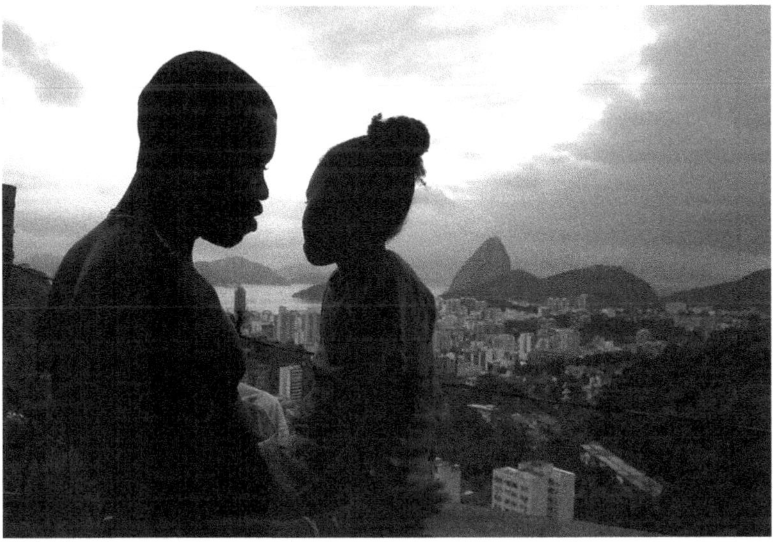

Figure 2.2 People at the favela of Pereira da Silva (*Pereirão*), Cirlan and his nephew
Source: Photo Sandra D'Angelo, Rio de Janeiro, 2010.

Two concepts can be particularly useful in describing the complexity of the contrasting emotions people feel towards the sense of place generated by baile funk: topophilia (from the Greek topos, 'place', and -philia, 'love') and topophobia (from the Greek topos, 'place', and -phobia, 'fear'). The terms later appeared in the work of the humanistic geographer Yi-Fu Tuan, who describes topophilia as the affective bond between people and place or setting (Tuan 1974: 4). The term 'topophilic' captures the way funkeiros positively engage with baile funk, while 'topophobic' describes the attitude of the part of Brazilian society, which perceives and represents the baile funk as a negative place where *funkeiros* have antisocial and 'dangerously provocative' attitudes and which is 'completely alienated [...] from any notion of social transformation or reality' (Pardue 2008: 35).

Baile funk, in contrast to samba, rejects the image of the Brazilian as a 'cordial' and 'industrious' mestiço, suggesting instead a new sense of Brazilianness exported abroad through baile funk, that I shall discuss in due course. This in turn, relies on what I perceive to be a postmodern 'loss of the real'. In baile funk, it is a disconnection with the local reality or postmodern 'loss of the real', which provokes topophobic reactions to the genre. Apart from a few examples of baile funk artists who choose to denounce their harsh life in the favelas through their lyrics, baile funk is currently more oriented towards a rupture with the local reality, which should be understood as the claustrophobic reality of the favela life. This is a tendency that provokes topophobia not only in part of Brazilian society, but also among practitioners of other popular music genres. The relationship between hip hop and baile funk which share a similar musical legacy illustrates this point. Hip hop in Rio de Janeiro was very similar to baile funk as Pardue discusses in his book' *Ideologies of Marginality in Brazilian Hip Hop* (2008): 'until 1998, "rap" and "hip hop" in Rio de Janeiro were basically indistinguishable from [baile] "funk"', but from late 1980s São Paulo hip hoppers excluded Rio de Janeiro from hip hop's circuit. '[From late 1980s] São Paulo hip hoppers generally excluded Rio de Janeiro from any sort of hip hop *imago mundi* because for them Rio 'rap' or 'funk' was completely alienated (*alienado*) from any notion of social transformation or reality' (Pardue 2008: 35). Illustrating hip hop's different approach to reality, Pardue explains that: 'In Brazil, hip hop is a form of identity expressed through the politics and pleasure of marginalized youth [...] hip hoppers represent the world as "reality" and "truth" in extremely dogmatic ways' (Pardue 2008: 1, 3). In this context *the real,* in hip hop, is represented by engagement with the urban social environment, a conceptual paradigm inherited from the Zulu nation philosophy of American DJ Africa Bambaataa, one of the originators of the hip hop genre in the 1980s (Pardue 2008: 108). As a result, hip hop's more committed approach to social issues became much more accepted in Brazil than baile funk which was seen as out of touch with reality in contrast. This tendency culminates in *funk de ostentação* (funk ostentation), one of the latest versions of the genre. Having emerged in São Paulo in 2011 (Freire 2012), *funk ostentação* is a subgenre of baile funk that it can be read as representative of the loss of a present reality in the way it promulgates an unrealistic social status. In *funk ostentação,* funkeiros are represented as buying into the tropes of conspicuous consumption that relates to a stereotypical perception of material success in the USA or Europe delivered through video clips on MTV and *YouTube* by some North American hip hop artists. Funkeiros and more specifically *funk ostentação,* can be said to lose touch with the *real* by proclaiming, through lyrics and music videos diffused on *YouTube,* the consumer-oriented lifestyle of capitalism. As Freire points out in her study about *funk ostentação,* all goods displayed in the *funk ostentação's* videos are fake or borrowed. What artists promote is not their reality (such as some North American hip hop artists for example) but is an invented one. The *funk ostentação* artist Mc Guimê declares in his interview with Freire (2012) that what is shown in videos is not their reality, including MCs displaying expensive clocks or jewellery; everything is not for real; it is not real, but just a reproduction (Freire 2012).

Additionally, baile funk can be also read as a postcolonial product in the process of developing its own way of 'writing back' to borrow the concept from *The Empire Writes Back* by Ashcroft (1989) about postcolonial literature and influencing the musical reality in the places from where baile funk absorbed musical features and cultural aspects through a transnational process of importing music. Yet, through the writing back process and the affirmation of baile funk internationally (Europe and North America), the favela reclaims its own place in the world. Through the process of relocation, baile funk music expresses a specific act that can be read as 'the right [of cultures or societies] to represent themselves by juxtaposing colonial inscriptions' (Fiedler 2007: 275). In baile funk, the process of postcolonial 'writing back' takes place through a participatory modality, through the relocation of music and place identity, transnationally from Brazil towards

Western EDM (electronic dance music) culture. As a demonstration of this process, it writes back in order to represent itself in the subgenre called 'Global or Transnational Bass', which is produced in all the main countries of Europe and the USA. Baile funk can be understood as the result of a global circulation of musical identities in the way it absorbs and reinterprets the sense of local and foreign place identity, relocated through music samples, from elsewhere. Thus, the most contemporary Brazil, the one that is permanently connected via the web (the majority of funkeiros and people living in favelas are connected via Internet and have a TV), is creating a valid alternative. An alternative that culminates in what Yudice observes in his essay *The Funkification of Rio* (1994), or in the 'funkification' of part of the European EDM in the way baile funk is 'funkificating' that is 'conquering space' in the work of independent and mainstream artists such as Diplo and M.I.A in Europe or Madonna, and the Black Eyed Peas in the USA. Since 2004, the year of production of the first successful song in baile funk style of the British artist M.I.A. and the American producer Diplo, the process of funkification of part of the European EDM culminated with the formation of new musical genres such as the 'Global Bass'. The genre incorporated the sense of place and a different sense of Brazilianness relocated by baile funk through which the poorest, youngest, and the most energetic Brazil legitimates itself. Baile funk, as a postcolonial musical product writes back to Europe exporting a different *self* and a different sense of place identity in contrast to the one planned and promulgated by the nationalistic inscriptions of the past.

A major criticism of baile funk music, in addition to those already exposed, is the sampling process, which is accused of dissolving the boundaries between human and machine-generated music, a practice decried by musicians who defend the beauty of real Brazilian popular music and the mastery of instrumental musicians. Rather than removing human agency, it would be argued that baile funk practitioners use these machines as a musical tool with extensive and precise programming capabilities allowing them to 'achieve specific and varying degrees of what they consider to be "human" musical expression' (Rodgers 2003: 316) The sampling practice itself is an ideal way to absorb the sonic environment of other places, where reinterpretation depends on a specific technical knowledge and a high presence of 'human agency' in the music production process. In the process of acquisition of electronic musical instruments and the adoption of foreign music, Carlos

Machado, aka DJ Nazz, made a major contribution to the baile funk genre. One of the most important DJs in the 1980s, playing every weekend at baile funk dances for around 5,000 people, Machado grew up in a suburb of Rio de Janeiro. He was determined to advocate the relocation of North American music from the US to Rio and the democratisation of samplers in Rio. As he himself states:

> I enjoyed the US. Sometimes I used to rent a car just to explore new places, driving two or three hours to discover new music and new places to find records. I used to go to different places: Rio to Miami, Miami to NY, NY to Los Angeles, Los Angeles to San Francisco, San Francisco to NY, NY to Rio de Janeiro. (quoted in D'Angelo 2014)

His topophilic attitude and his active networking with people in the North American music business (Figure 2.3), musicians, music store

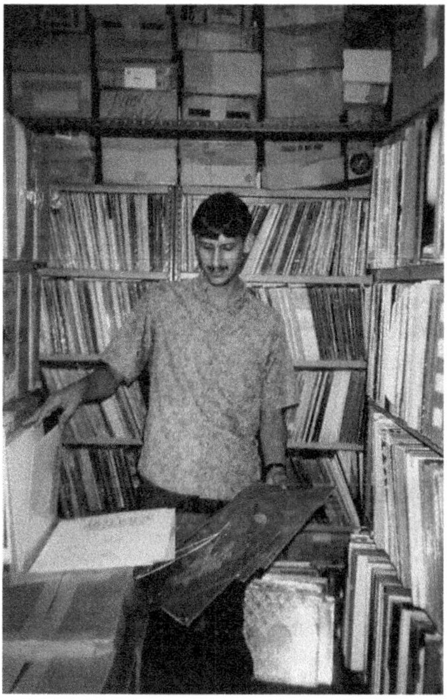

Figure 2.3 DJ Nazz (Carlos Machado) in New York
Source: Photo courtesy of Carlos Machado, New York, 1983.

owners, and radio and label representatives had significant repercussions in terms of the development of the process of adoption of foreign international musical trends in Rio de Janeiro. Nazz was operating away from the official channels of music distribution in Brazil and thanks to him Brazilian musicians and party organisers had access to cheaper musical technology instruments. As DJ Nazz states:

> The Akai [a sampler] was still very expensive; by that time the cost was around 4000 USD, Numark 1000 UDS, and Gemini which was around 300 USD was much more affordable for the *equipes de som* [sound systems]. [...]
>
> The first mixer with a sampler, the Numark model 1775, was brought by me. This mixer allows the DJs to sample small portions of songs and start to do live medleys[6] that *funkeiros* call *montagem*. Then I started importing Gemini from the US, that democratized the access to music technology. (quoted in D'Angelo 2014)

Hence, the genre of baile funk can be read as postmodern in the way it considers technologies which do not just serve to preserve and transmit music, but are necessary for its creation. In the next section, I will present some of the findings of my empirical research on the impact of sampling and music samples in the relocation of the sense of place.

Relocation of music, the use of samples and sound effects in Baile Funk

Going back to the discourse surrounding sampling art and the selection of samples, digital technology has facilitated a greater degree of participation among DJs who are able to operate within the transnational flow of music. An example of their incorporation of the sense of foreign places into baile funk music is a popular fragment of sound extracted from a traditional southern Italian musical form, the 'tarantella'. There are several types of tarantella in southern Italy. Through a comparative method of several fragments of sounds from where I recognised a piano accordion accompanied by a tambourine, characterised by a compound upbeat tempo, usually in 6/8 time (sometimes 18/8 or 4/4) I evinced that the one incorporated in baile funk is a Neapolitan tarantella, the *tarantella napoletana*. DJ producers have been incorporating into this sample an uncountable number of baile funk songs, such as *Abertura Funk*, *Chapadona de Skol* by DJ Byano and *Aquecimento Italiano* by DJ Muleke, in live medleys since early 1990s.

The Brazilian musicologist Oneyda Alvarenga (1953) describes the role of Italian culture in the historical development of Brazilian literature and music, such as work by Buarque De Holanda (1902–1982), who focuses on the multiple cultural influences that forged twentieth-century Brazil in his book, *Roots of Brazil* (1936) and describes the considerable contribution of Italian culture to Brazilian development. There was considerable Italian immigration to Brazil, where, according to the Italian government, there are 25 million people of Italian descent, the largest community outside of Italy itself (Itenet 2007). Through a direct observation during my fieldwork, I noticed that Italian traditional musical heritage is part of the soundscape in Brazilian mass media culture such as soundtracks in Brazilian popular telenovelas. One of these is called *Terra nostra* ('Our land' in Italian), which has as a main topic the *saudade* (nostalgia) for the lost country, a term and a concept usually employed to express nostalgia for Africa. In this case, the sample taken into account delivers a sense of legacy and nostalgia for Europe and Italy. This sample can be read as an affirmation of the Brazilian historical inheritance which through a semiotic sign delivers the semantic aspect of a topophilic attitude towards the sense of place delivered, in this case, by this type of traditional popular music from southern Italy.

Let us now to move to how a sense of place can be evoked through the employment of sonic fragments typical of a specific social soundscape. '"Parapapapapapapa/ Paparapapapapapa/ paraapapapapapa kla que bum/ parapapapapa"; the rhythmic onomatopoeic lyrics refrain of the song "Rap das Armas" (Rap of Weapons) imitates the discharge of an automatic gun' (Tiis 2010: 1). We can understand the sense of place delivered by this sample in two opposing semantic aspects. First as a representational sign in the way it can be understood through the lyrics of the song, or it can be read in an opposite semantic order that dissolves the re-experiencing/re-evoking a soundscape indicative of social tension. In this way it exteriorises a desire for potency and a sense of abundance while euphorically celebrating the affirmation of baile funk's utopian dimension and sense of place. Furthermore, besides understanding gunshot as a semiotic representational sign we can further evince on its postmodern aspect, in the way the soundscape of a conflicted present is congruently juxtaposed with a semantic representation of soundscapes of the past. Consequently, we can read this sample as a semiotic sign which encapsulates and delivers the memory and a sense of the contemporary soundscape as well as the memory of Brazilian musical heritage and the mass media soundscape of the past.

I argue that the gunshot was extracted from a song very popular during the 1960s, which was part of the *sertaneja* (from the countryside) genre.[7] The songs *O Homem Mau* (*The Bad Man*) and *O Rock Bravo Chegou Para Matar*[8] (*The Angry Rock Came to Kill*), from which the sample was taken, were recorded in the late 1960s and produced by the duo Leo Canhoto and Robertinho. *O Rock Bravo Chegou Para Matar* was inspired by 'bang-bang *italiano*' movies, a style popular in the 1970s. The peculiar feature of bang-bang italiano was its use of sound effects such as the gunshot (*bang-bang!*) and dialogues/speech (*diálogos*) between people or between policemen and criminals. This sample appears in baile funk tracks such as *Toma Toma!* (*Take This! Take This!*) and *O Meu Bonde Tá Passando* (*The Crew Is Coming*) by MC Sabrina, as well as many other successful tracks produced by MC Marcely, MC Maiquinho and DJ LP de Friburgo.

Soundscape researcher Järviluoma termed this re-experiencing of a sonic event 'hearing-point memory', a term highlighting the connection between the sense of place of the event and the emotional state of the person at the time of recall. As Järviluoma argues, sounds and memories are not just copies of the past in the present, but are always in transition (Järviluoma 2009). In this sense, memories, like sound, are thus both contextual and temporal and Järviluoma goes on to argue that aural memories are connected with the context or place where they were created. To re-evoke a place and its sense of place, funkeiros evoke the acoustics of that place. The past and the way it is remembered is coloured by one's current situation and state of mind and we could extend this concept to argue that one's state of mind is similarly influenced by the soundscape and the place in which the sound is heard. The state of mind changes because of the music and what it recalls. In the case of baile funk, the utopic place discussed by Sneed (2003, 2008) is reached through a transformation of the state of mind of people who react to acoustic memories.

All baile funk tracks tend to be built upon a pictorial spatiality. To do so, paramusical acoustic spatial indicators such as reverberation and 'big echo' effects are applied on samples of voices, singing or fragments of sounds or melodies. In the literature, the semantic aspects of sonic variables, which can be defined as 'indexical signs' (where the signified causes the signifier, in the same way that smoke signifies fire), are thus uncovered 'An echo and reverb could alter the emotive impact of sounds produced, and the affective change often [seems] to be out of all proportion to the purely sonic changes wrought by the effect' (Doyle 2005: 4); while Rodgers asserts, 'Analysis of how a particular grain or texture is constructed during the sampling process – through techniques of sound design, spatialisation, panning, and effects processing – can also offer insight into how samples resonate with particular musical functions and cultural meanings' (2003: 317). Echo and reverb are qualities of *real* physical spaces, the creative action here connoting a subversion of the 'indexical signs' into a postmodern loss of the real. A sample of a dialogue between a policeman and a gangster extracted from a song from a traditional genre, the sertaneja, called *Homen Mau* and produced in the 1960s, demonstrates a typically extensive application of reverb. The reverb applied here functions to reshape the spatio-acoustic conditions, rendering a sense of power, which amplifies and affirms an uncanny representation of the self in place. In dialogue with Doyle and Rodgers, we can say that 'with the addition of echo and reverb, place and space had become part of the larger musical equation, a new component in the musical tonality' (Doyle 2005: 5, 6).

Conclusions

In Brazil, baile funk struggles for acceptance, but abroad this genre is positively welcomed (Miranda 2012; Sá 2007). The relocation of baile funk has been powered by the impressive effort of DJs, promoters, and record labels such as Man recordings in Berlin, the club 'favela chic' in Paris, Irma records in Bologna or Mr Bongo in London. The EDM community has thus facilitated the spread of a new idea of Brazil abroad. However, alongside baile funk music, other factors have contributed to the development of a different perception of Brazil. Also important have been the Internet, and some contemporary Brazilian films such as *City of God* or *Tropa de Elite*. Together, these developments have presented a sense of Brazilianness artificially constructed and spread by political strategies of the past. One of the consequences of this transnational journey is the creation of a new musical genres in Europe, the so-called Transnational or Global Bass, a recent type of music based on baile funk samples. Above all, my aim in advocating the study of baile funk's sense of place within this fascinating, multilayered culture, is to keep open a dialogue and to provide a counterpoint to topophobia as the alleged consequence of a postmodern loss of the real. To conclude, this study opens the way for further research into how baile funk music became a powerful iconic poetic strategy for many youths to create and spread their 'sense of place' within the contemporary transnational flow and circulation of sounds.

Notes

1. From 1960 and 1970, a conspicuous number of *bailes* (dances) and *bailes black* (black dances) were organized by sound systems in Rio de Janeiro. They were an important form of leisure for people living in the poorest areas. These dances were also fundamental in the way they permitted the musical and social development to come into formation from the late 1980s under the baile funk.
2. In Brazilian Portuguese 'baile funk' can be literally translated as 'funk dances'. The term 'funk' derives from the North American funk music played at dances in Rio de Janeiro during the1970s. Although, in the 1980s the music played at those dances radically changed, and shifted from North American funk to a sample-based type of electronic music, the name 'baile funk' remained invariate. In Brazil, baile funk is currently used to indicate both a dance and a specific type of Brazilian electronic dance music. Internationally, baile funk indicates only the genre. In this chapter, I refer to the term baile funk as a genre as it is currently understood among European musicians.
3. *Baile black* appears in Giacomini, Sonia Maria (2006). *A alma da festa: família, etnicidade e projetos num clube social da Zona Norte do Rio de Janeiro, o Renascença Clube* (Belo Horizonte and Rio de Janeiro: UFMG, IUPERJ), pp. 189–256. Thayer, Allen (2006). 'Brazilian Soul and DJ Culture's lost chapter', *Wax Poetics* 16, pp. 88–106. Essinger, Silvio (2005). *Batidão: uma história do funk* (Rio de Janeiro and São Paulo: Record), pp. 15–48. McCann, Bryan (2002). 'Black pau: Uncovering the history of Brazilian soul', *Journal of Popular Music Studies* 14, pp. 48–54. Frias, Lena. 'Black Rio: o orgulho (importado) de ser negro no Brasil', *Jornal do Brasil*, Caderno B, 17 July 1976, 1 and 4–6. Online at http://festablax.multiply.com/photos/album/24/Ma teria_Black_Rio_de_Lena_Frias_- _Jornal_do_Brasil_170776_(accessed 08 February 2008).
4. For a more details about the use of the term *funk* in Brazil see Livio Sansone (2003: 114).
5. See http://www.ssrede.pro.br/ibge%20estatisticas.pdf (accessed 8 August 2014).
6. In music, a medley is a piece composed from parts of existing pieces, usually three, played one after another, sometimes overlapping. They are common in popular music, and most medleys are a mix of songs rather than instrumental music. A medley which is a remixed series is called a megamix; this is often done with tracks for a single artist, or with popular songs from a given year or genre.
7. The *musica sertaneja* genre originated in the 1920s in the *sertão* (countryside), far away from Brazilian cities (*Dicionário Cravo Albin da Música Popular Brasileira*, 2012).
8. A song from the album *O Rock Bravo chegou para matar* (RCA 1970).

Works cited

Alvarenga, Oneyda (1953). *Musica popolare brasiliana* (Milano: Sperling & Kupfer).
Ashcroft, Bill (1989). *The Empire Writes Back: Theory and Practice in Post-colonial Literature* (London and New York: Routledge).

Bennett, Andy (2004). 'Music, space and place', in Whitley Sheila, Andy Bennett and Stan Hawkins (eds), *Space and Place: Popular Music and Cultural Identity* (Farnham: Ashgate), pp. 2–8.
De Holanda, Sergio Buarque (1936). *Roots of Brazil* (Notre Dame, IN: University of Notre Dame Press).
Dennison, Stephanie (2005). 'Favelas', in Lisa Shaw and Stephanie Dennison, (eds), *Pop Culture Latin America! Media Arts and Lifestyle*. pp. 350–353
Doyle, Peter (2005). *Echo & Reverb: Fabricating Space in Popular Music Recording 1900–1960* (Middletown: Wesleyan University Press).
Dyer, Richard (1999). 'Entertainment and Utopia', in Simon During (ed.), *The Cultural Studies Reader*, 2nd ed. (London: Routledge), pp. 371–81.
Fiedler, Matthias (2007). 'Belated memory: The "postcolonial eye" on Africa in Uwe Timm's novel *Morenga*', in Aileen Pearson-Evans and Angela Leahy (eds), *Intercultural Spaces: Language, Culture, Identity* (New York: Peter Lang Publishing).
Freire, Lybni Silvia (2012). 'Nem luxo, nem lixo: Um olhar sobre o funk da ostentação' IX POSCOM Seminário dos Alunos de Pós-Graduação em Comunicação Social da PUC-Rio 07, 08 e 09 November 2012.
Itenet Rapporto degli Osservatori Itenets Regionali Anno (2007). http://www.esteri.it/mae/doc_osservatorio/rapporto_osservatori_2007.pdf (accessed April 2013).
Järviluoma, Helmi (2009). 'Soundscape and social memory in Skruv', in Helmi Järviluoma, Meri Kyto, Barry Truax, Heikki Uimonen and Noora Vikman (eds), *Acoustic Environments in Change* (TAMK University of Applied Sciences; University of Joensuu, Tampere: Tampereen ammattikorkeakoulu).
Miranda, Gabriela (2012). 'The appropriation of technologies in circuit funk carioca', MA thesis (Rio de Janeiro: PPGCOM-UFF, Universidade Federal Fluminense).
Neuwirth, Robert (2005). *Shadow Cities: A Billion Squatters, a New Urban World* (New York: Routledge).
Ovalle, Priscilla Pena (2011). *Dance and the Hollywood Latina: Race, Sex and Stardom* (New Brunswick, NJ: Rutgers University Press).
Palombini, Carlos (2010). 'Notes on the historiography of *Música Soul* and *Funk Carioca*', *HAOL* 23, pp. 99–106.
Pardue, Derek (2008). *Ideologies of Marginality in Brazilian Hip Hop* (New York: Palgrave Macmillan).
Perrone, Charles, and Christopher Dunn (2001). 'Chiclete com Banana', in Charles Perrone and Christopher Dunn (eds), *Brazilian Popular Music and Globalization* (University Press of Florida).
Reily, Ana Suzel (2000). 'Introduction: Brazilian musics, Brazilian identities', *British Journal of Ethnomusicology* 9(1), pp. 1–10.
Relph, Edward (1976). *Place and Placelessness* (London: Pion).
Rodgers, Tara (2003). 'On the process and aesthetics of sampling in electronic music production', *Organised Sound* 8(3), pp. 313–320.
Sá, Simone (2007). 'Funk carioca: música popular eletrônica brasileira?!', *Revista E-Compós* 10. http://www.compos.org.br/seer/index.php/e-compos/article/view/195 (accessed February 2011).
Sansone Livio (2003). *Blackness Without Ethnicity: Constructing Race in Brazil*. (Houndmills and New York: Palgrave).

Shaw, Lisa (1998). 'Coisas Nossas: Samba and identity in the Vargas Era (1930–1945)', *Portuguese Studies* 14, pp. 152–69.
Sneed, Paul (2003). 'Machine gun voices: Bandits, favelas, and Utopia in Brazilian funk', PhD dissertation (University of Wisconsin-Madison).
Sneed, Paul (2008). 'Favela Utopias the *Bailes Funk* in Rio's crisis of social exclusion and violence', *Latin American Research Review* 43(2), pp. 58–79.
Tiis, Elena (2010). 'Of musics and bodies: Embodying the Brazilian favela funk', *Noise and Capitalism*. http://blogs.arteleku.net/noise_capitalism/?tag=elena-tiis&lang=es (accessed August 2014).
Tuan, Yi-Fu (1974). *Topophilia: A Study of Environmental Perception, Attitudes, and Values* (New York: Columbia University Press).
Tuan, Yi-Fu (1977). *Space and Place: The Perspective of Experience* (Minneapolis: University of Minnesota Press).
Tuan, Yi-Fu (1980). 'Rootedness versus sense of place', *Landscape* 24(1), pp. 3–8.
Yudice, George (1994). 'The funkification of Rio', in Tricia Rose and Andrew Ross (eds), *Microphone Fiends: Youth Music and Youth Culture* (New York: Routledge), pp. 193–217.

Interviews

D'Angelo, Sandra (2014). Interview with DJ Nazz, Rio de Janeiro, 12 February 2011.

3
Die Antwoord: The Answer to the Unspoken Question

Katarzyna Chruszczewska

A South African group Die Antwoord, which in Afrikaans means 'the answer', began their international career almost immediately after the recording of their first album *O* (2009). The music they perform, defined by them as '*zef* style', is a mix of rave and rap juxtaposed with the overwhelming visuality of their videos.[1] The phenomenon of their rapid popularity seems to be the result of a striking contrast provided by their street music and the sophisticated visual *brut* aesthetics that is inspired by the most prominent contemporary African artists.

Although various scholars have discussed the popularity of Die Antwoord (Haupt 2012b; Krueger 2012; Marx and Milton 2011; du Preez 2011; Truscott 2011), the visual aspect of their performance, crucial for my argument, seems to be neglected, even though it plays an important role in understanding the Die Antwoord phenomenon. In this chapter, I plan to use Die Antwoord as a case study to analyse the means and purposes by which popular music negotiates identity and place: hybridising what is global and local, familiar and exotic. The analysis of the Die Antwoord's phenomenon invites discussion of local issues in the context of global hip hop culture, particularly the unusual conjointment of Western and South African aesthetics. In my analysis of the complex relations between South African and Western identity depicted in the music videos of the group, I will argue that the main focus of Die Antwoord's art is to re-establish those relations through the relocation of South African culture.

The key notion that defines Die Antwoord's style is *zef*, a uniquely South African concept, derived from the Ford Zephyr car symbolising the South African working class of the 1950s and 1960s. It connotes a 'white-trash' Afrikaner aesthetic as well as an Afrikaners' music genre. Yet, *zef* has ironically distanced itself from what it originally represented.

While it previously represented what is 'poor' and 'common', its meaning has shifted toward 'fresh' and 'cool' (Truscott 2011: 97). It is impossible to decide whether Die Antwoord's endeavour remains a parody of *zef* or – as a parody of a parody, which defies a fixed meaning – its perfect realisation. What is certain, however, is that an element of parody is present in Die Antwoord's performance. The reason behind this strategy might be an attempt to distance South Africa from its difficult past by transforming post-apartheid identity (Truscott 2011) into an intertwined, multicultural nationalism, celebrating the hybridity of South African culture (Marx and Milton 2011). What interests me is the overall aim of Die Antwoord's parody and the particular strategies adopted by the band of reconfiguring the space and remapping the body. Hence, I will focus on Die Antwoord's attempt to construct a hybrid identity that criticises Western mainstream culture and relocates hip hop into the South African music genre *zef*.

Hip hop reimagining and remapping the space

In order to place Die Antwoord on the map of global hip hop culture,[2] it is essential to briefly present what constitutes hip hop culture itself. It would be grossly inaccurate to define hip hop as a music genre. The term 'hip hop' describes a particular lifestyle as well as a subversive and resistant stance towards social subordination and marginalisation. It encompasses various elements: rapping, mc-ing, freestyling, break dancing, and graffiti (Rose 1994: 34–61); central notions of hip hop music expression include break, flow, sampling, and mixing (Terkourafi 2010: 4–6). Due to tension between the local and the global, and the quest for authenticity, hip hop cannot be conceived as a monolithitic phenomenon. It can be seen as a genre in the process of continuous hybridisation, open to a variety of local and ethnic influences. 'With sampling and mixing at its very core, hip hop provides an expressive vehicle that is flexible enough to accommodate the multiple origins and concerns of increasingly interconnected and mobile populations', observes Terkourafi (ibid.: 5). Moreover, local variations of hip hop remain in a constant dialogue with the global hip hop culture.

Hip hop can also be classified as a discursive practice. It forms a frame for reinterpretation of social knowledge and cultural practices, tackling such issues as gender and race, redefining blackness and whiteness. It provides opportunities to contextualise and negotiate identities operating through means of musical and verbal discourse as well as visual representations (Connell and Gibson 2001: 3). Particular hip hop linguistic practices can involve 'language crossing': transgressing social boundaries by conscious codeswitching and hybridising of language (Androutsopoulos 2009: 44–46). Its aim is to transform the identity of its users, negotiate a subject position between the local and the global, the privileged and the discriminated. It 'expresses a form of politics perfectly suited to the postcolonial era' (Lipsitz 1994: 36), as a major source of the genre's content is centred around issues of ethnicity and power.

It is widely acknowledged that music is not limited to the representation of a particular time and place. Due to the transformative creativity of hip hop artists – mixing and sampling, or, in other words, cultural *bricolage* (Bennett 2001: 90) – it is possible to re-imagine and re-map local space, and, what follows, deconstruct the space of the hostile, post-industrial city and reconstruct it on completely different terms (Rose 1994: 34). Hence, rappers are able to function as alternative cartographers, reconfiguring identities and examining spatial partitioning of race (Forman 2004: 202). The emergence of the first rap scenes in New York, Miami, Houston, Seattle, San Francisco, and Los Angeles, remapped cultural influences by dividing the space between the East and West Coast. It seems that dissemination of the hip hop phenomenon has played a prominent role in a passage from sociopolitical invisibility to cultural presence.

As hip hop is based on spatial logic, it provokes the discussion on place. However, it itself could be seen as a space, in which identities are being negotiated (Omoniyi 2009: 114). In my analysis of Die Antwoord, I will focus on their performance as a space for the negotiation of place and identity through the means of cultural *bricolage*: visual sampling and introduction of the *zef* genre, thereby forming a hybrid of local and global influences. I will argue that through their hip hop narration, Die Antwoord attempts to re-configure South African culture, reimagining the urban- and suburbanscape and reinventing post-apartheid identities.

Die Antwoord and the *Zef* side

The immense popularity of Die Antwoord seems to result from the startling combination of rap-rave music in conjunction with visually exciting and provocative music videos. The significance of linking the aural and the visual arises from the growing importance of new media to the process of dissemination of music. Such major internet platforms as MySpace, SoundCloud and, above all, YouTube have radically changed the process of discovering music. YouTube, following and eventually outdoing the success of Music Television, has continued to influence the very experience of music, intimately connecting sound with vision.

Die Antwoord was formed in 2008 as an underground hip hop collective and despite the unconventional character of their performance, their international career began almost immediately after recording their first album *O* in the following year. Die Antwoord's first hit was the single *Enter the Ninja* (2010). The accompanying video directed by African cinematographer, Rob Malpage reached over 23 millions views on YouTube, while the group's biggest hit *I Fink U Freeky* (2012) got over 52 millions views. Others, 'less popular' songs reached between 10 millions and 20 millions, including *Baby's on Fire* (2012), *Fatty Boom Boom* (2012), or more recently *Cookie Thumper* (2014). Additionally, Die Antwoord engaged in the production of several mock-documentaries, of which the most famous, *Zef Side*, a short movie by Sean Metelerkamp (2009), reached over 13 millions on YouTube and contributed significantly to the group's popularity. The members of Die Antwoord, Yo-landi and Ninja, also starred in the narrative short film *Umshini Wam* (2011), directed by Harmony Korine, known for the controversial *Kids* (1995) and his latest movie *Spring Breakers* (2013). Their participation in mock-documentaries, short fiction films, street-art projects and other artistic endeavours provokes thinking about the group in terms of artistic performance rather than solely in terms of music production.

The movie *Zef Side* as well as following projects such as *Straight from the Horse's Piel*, reinforces the significance of the music's particular sites of origin, referring to spatial discourse of hip hop which emphasises the importance of such notions as 'ghetto' or 'hood'. In the picture, Yo-landi Vi$$er and Ninja are presented on a background of poor suburban areas of Cape Town, within which they tell stories regarding work on their rapping with the aid of DJ Hi-Tek, who 'owns a PC computer and makes next-level beats' (*Zef Side* 2009). This strategy operates with space that is 'simultaneously real, imaginary, symbolic and mythical' (Forman 2004: 217). It links *zef* music to that space: re-imagined urban areas of Cape Town, and sets the stage for further performance. The concept of *zef* defines the manner in which the space is re-invented and consequently depicted in Metelerkamp's movie. Identically to *zef* itself, *Zef Side* glamourises poverty by referring to the anti-aesthetic: seeking diversion in ugliness and decay.

In *Zef Side*, the suburban area of Cape Town is characterised by its inhabitants: women with their weary faces, toothless and maimed men. The place has a mythical character removed from time and space. With endless lines of identical dowdy houses, separated from the rest of the city by the concrete fences, it seems to escape urban spatial organisation and the current of time. Yo-landi, Ninja, and DJ Hi-Tek have lived there since they were born, or according to Yo-landi 'since forever' (*Zef Side* 2009), and hence are presented as intrinsic elements of the poor Cape Town district. Like the space around them, they are at the same time vulgar and nuanced, explicit and intriguing. All three are either depicted leaning against the concrete fence in the static documentary-like interview, or dancing in the front yard cluttered with old buckets, broken plant bowls, a rusty bicycle, and car doors. The latter is filmed in the music video style, using close captures, slow motion, and saturated colours. Metelerkamp presents a phantasmagorical, perhaps even nostalgic, image of the forgotten poor Cape Town suburbs, which is at the same time disquieting and dreamy. It may be read as a symbol of South African urban peripheries and cultural peripheries in general.

The image deriving from the *Zef Side* is of working-class 'white-trash' Afrikaners, of which Die Antwoord appears entirely representative. The most prominent band members, however, are the well-educated and well-resourced duo, Watkin Tudor Jones and Anri du Toit who adopted the working-class personas of Ninja and Yo-landi. Offstage husband and wife, they have collaborated before in numerous projects, including: Max Normal, The Constructus Corporation, and Max Normal.tv. In each of them, both Jones and du Toit took up a different persona that often reflected a particular class of the South African society. Max Normal, for instance, was a typical representative of the middle class, while Watkin Tudor Jones (as a solo project), who always dressed in white and often held a golf club, referred clearly to those of a higher social class. The artistic strategy of Watkin Tudor Jones, who adopts various personas and thus operates with different stereotypical representations, seems to mock the compromised social stratification of South African society. The critical character of this performance reveals itself in multiplication of clichés regarding particular social strata. In the case of Die Antwoord, the oversimplified image of the working class is intensified by exaggerated use of stereotypical connotations for 'white trash', such as vulgarity, obscenity, and violence. Thus, Die Antwoord's performance becomes a grotesque representation attacking both the post-apartheid society with its prejudices and the mainstream culture, which preserves such stereotypes.

Officially, Die Antwoord consists of three members: rappers Yo-landi Vi$$er (Anri du Toit), and Ninja (Watkin Tudor Jones), as well as DJ Hi-Tek, a more elusive character, who seems rather a virtual than an actual person. Portrayed by various men he is most often absent during interviews, but the function of his presence in the group offers potential for a number of musical contributors. The vivid personas adopted by the members of the band remain a significant element of Die Antwoord's phenomenon, relocating identities based on class and race.

The persona of Yo-landi Vi$$er recalls the stereotypical image of femininity presented in rap songs and rap videos: sexual object, on the one hand, and a promiscuous and materialistic woman, on the other. Remaining a 'Rich Bitch', as claims the title of one of Die Antwoord's songs, which is sung almost exclusively by Yo-landi, is not the only characteristic of her identity. Yo-landi's persona is both extremely sexual (*Baby's on Fire* 2012) and violent (*Fatty Boom Boom* 2012). While embracing promiscuity and violence with grotesque exaggeration, she seems to satirise the stereotypical gender roles provided by the mainstream rap culture. Yo-landi challenges binary oppositions, remaining at the same time beautiful and scary, minor and elder. Because of her diminutive body features and tiny voice, Yo-landi resembles a minor, albeit intensely sexualised (*Enter the Ninja* 2010). The combination of eroticised violence and child-like features makes her persona unsettling, yet intriguing. She seems to recall an ageless demon woman extremely tempting and dangerous; a mythical creature from the African imaginary, as, for example, a voodoo daemon, which the band refers to in *Fatty Boom Boom* (2012). Yo-landi's persona, simultaneously connoting Cape Town marginality, rap culture, and African imaginary, is overloaded with meaning, which results in the implosion of sense. What seems to lie beneath the series of masks adopted by Yo-landi is not a fixed identity, only a chaotic vortex of cultural influences.

Accompanying Yo-landi is the rapper Ninja, a persona created by Tudor Jones. In the media, Tudor Jones has not yet abandoned the raw, violent, and naïve character of Ninja, and eventually provoked the question regarding his previous projects and personas. 'Ninja is, how can I say, like Superman is to Clark Kent. The only difference is, I don't take off this f**ken Superman suit', explained Tudor Jones in *Straight from the Horse's Piel* (2010). Ninja mimics, hairstyle, golden teeth, body covered with low-quality tattoos which all refer to the mixed-race urban low-class community. Tudor Jones invokes the stereotypical image of the social margin in order to transform it into the space of rebellion. Recalling *zef*, he revaluates social roles so that the situation of an underclass becomes a position of power: a source of artistic creativity.

The section on the band's members from Die Antwoord's official website claims that 'Ninja iz by far da hardest, ugliest, most gangster-ass playa in da game right now' (Die Antwoord 2013). Through his exaggerated image, Ninja represents a parody of a certain kind of masculinity, mainly based on sexuality and violence (du Preez 2011: 104). What his character simultaneously connotes is 'white-trash' Afrikaners and the gangsta-rap tradition, white and black underclass. From the position of a social outcast, Ninja contests contemporary society and culture: 'fok julle naaiers' ('f**k you f**kers'), he repeatedly claims in the song under the same title (*Fok Julle Naaiers* 2012). The purpose of his performing a trash identity, based on aggression and contempt, is the disruption of the actuality defined by the apartheid past. What seems to provoke his attack are clear-cut categories and fixed social roles that impede society. Thus, rebelling against imposed identities, Die Antwoord celebrates social transgression and cultural contamination.

Ninja's contestation, however, goes beyond the verbal layer of Die Antwoord's performance. His body, covered with numerous tattoos, resembles the spatial organisation of Cape Town and thus resembles a map. Ninja's haircut, angry mimics, and manner of speaking indicate his alleged Afrikaans suburban background, while gang tattoos refer to the Cape Flats area inhabited mostly by black South Africans. There are at least three gang affiliations tattooed on Ninja's skin, as if his flesh, like a South African city, was divided between different zones of influence. Most of the tattoos are prison-inspired. They mirror power relations between prisoners, a disturbed reflection of the Cape Town's topography. The other 'street style tattoos' (*Take No Prisoners* 2010), read as follows: 'ugly on the skin, pretty from within', 'how can a (*sic*!) angel break my heart?', 'if you don't like funerals don't kick sand in ninjas face', and 'pretty wise'. There is also a knife tattoo, the image of Richie Rick and Die Antwoord's mascot, 'Evil boy': the combination of Casper, the friendly ghost, with an enormous penis. What is marked, hence, on Ninja's body are different cultural references to South African gangs, street wisdom, American imaginary, and rap culture. The intertwinement of cultural influences marked on Ninja's skin indicates the attempt to construct a transracial, South African, identity which is at odds with a contemporary society still influenced by post-apartheid order.

'Die Antwoord are a fre$ (*sic*!), futuristik, flame-throw-flow-freeking, *zef* rap-rave krew from da dark dangerous depths of Afrika', asserts the band's website (Die Antwoord 2013). The image of the group remains on the border of kitsch, obscenity, and the grotesque. Die Antwoord multiplies the clichés regarding hip hop artists and the South African underclass in order to expose the collapse of authenticity in Western mainstream culture. Their unique style results from the curious combination of African influences, hip hop, and popular culture, a juxtaposition of their raw *zef* music with the sophisticated aesthetic of their videos, which I examine in the following part.

Die Antwoord aesthetic

In hip hop culture, the sonic aspects of performativity are quite often accompanied by the visual aspects: fashion, various kinds of street art and particularly graffiti. The 'Highly visual and phantasmagoric aesthetic of the [hip hop] music and its culture' is mentioned by Murray (2004: 7), while Thompson investigates the mutual influence of hip hop and high art (2009). What is interesting here is the use of certain aesthetics to complicate the meaning of identity.

The elaborated visual aspects of Die Antwoord's performance is the effect of collage or, in hip hop terminology, of sampling of original sources, so that the achieved result is simultaneously something familiar and unexpected. Die Antwoord's sources of inspiration include the white Afrikaans culture, gangsta-rap, folk African culture with its demons and voodoo elements, and pop culture with Pokémon onesies and smiley faces. It is also possible to find references to African high art: the photography of David Golblatt, who documented apartheid among white-trash Afrikaners, the photography of Araminta De Clermont, who portrayed prison inked black gangsters of African cities and the disturbing photography of Robert Ballen, depicting deformation, filth, and madness. The direct references include, among others: the painting titled *Black Gynecologist* by Anton Kannemeyer (2008), or the sculpture *The Butcher Boys* by Jane Alexander (1985/86). Besides citing various artists and referring to culture images, Die Antwoord has also collaborated with several artists. I have already mentioned the performer Leon Botha and cinematographers: Rob Malpage, Sean Metelerkamp and Harmony Korine. They also collaborated with the upmarket New York fashion designer, Alexander Wang, who used the single *I Fink U Freeky* for the campaign of his *T line* and designed the 'voodoo' dress, used in the video of the song. Another important contributor is the aforementioned Ballen, a contemporary photographer of South Africa. He designs characteristic graffiti, which can often be seen in the background of Die Antwoord's videos. He was also the director of *I Fink U Freeky* (2012), which consists of a series of moving images that refer to colonial associations of Africa with filth, horror, abjection, and cruelty. In his black-and-white photography of South Africa, Ballen focuses on the excluded, as in his *Asylum* series (Ballen 2014), or *Shadow Chamber* (Ballen 2005). Photographs of the latter are shot in his studio, a shabby room with disturbing drawings on the wall and doll parts on the floor. The visual discourse of race, apartheid, and colonialist past, presented particularly in the examples of Cape Town and Johannesburg, is combined with visual junk served by the Western mainstream culture. These carelessly crafted Die Antwoord videos enhance the notion of inauthenticity of their personas. While the collaboration with renowned South African artists incorporates Die Antwoord's performance into high art, discussing critical social issues, such as discrimination, separation, and exclusion, the contribution of, for instance, Alexander Wang, suggests the commercial edge of their project.

Die Antwoord's videos are set either in a dark, mythic-like space (*Enter the Ninja* 2010; *Evil Boy* 2010; *Fok Julle Naaiers* 2012; *I Fink U Freeky* 2012), or in the Johannesburg cityscape (*Baby's on Fire* 2012; *Cookie Thumper* 2013; *Fatty Boom Boom* 2012; *Pitbull Terrier* 2014). In both cases, the background of the video includes characteristic graffiti paintings inspired by Roger Ballen. All Johannesburg's videos are set in a different part of the city. It is, respectively, a mawkish middle-class house in *Baby's on Fire*, a street corner of the inner city in *Fatty Boom Boom*, a poor underclass area in *Cookie Thumper* and abandoned construction site in *Pitbull Terrier*. With real urban sites being transformed into spaces of art carefully framed and often transformed by graffiti paintings, Johannesburg becomes a palimpsest of real and fake, local and global, kitsch and art. The change in the urbanscape may result from the new form of orientalism, namely the attempt to meet the expectations of a Western audience regarding South African image. Hence, it would be a part of the elaborated strategy to market the cultural margins. On the other hand, it may well be the effect of launching the discourse of trauma that *brut art* aesthetic of graffiti figures suggests. The transformation of Johannesburg's cityscape marks the return of the repressed, colonial, and apartheid past in the psychoanalytic work of uncovering the hidden. The result is uncomfortable familiarity, evoking the Freudian uncanny (Freud 1919). The sophisticated visual layer of their performance is contrasted with Die Antwoord's musical style which is coarse and aggressive. Even though Die Antwoord communicates with its audience using the familiar language of hop-hop culture, due to its dark aesthetic their performance appears exotic and weird.

Between the local and the global: identity and place

The hybridity of aesthetic influences could be understood as a form of subversion against imposed fixed identities. It deliberately disrupts homogeneity, exploring 'in-between' spaces of postcolonial culture. Hybridity has the empowering potential, since it 'reviews the past, refiguring it as a contingent "in-between" space, that innovates and interrupts the performance of the present' (Bhabha 1994: 7), in which identities can be negotiated and bastardised. What is at stake here is 'the undermining not only of previously clear-cut categories of race, ethnic background, national affiliation, and so on, but also of the certainties that emanate from these distinctions' (Terkourafi 2010: 6).

The opening passage from Die Antwoord's debut single read as follows: 'I represent South African culture ... a lot of different things ... blacks, whites, coloureds, English, Afrikaans, Xhosa, Zulu – *watookal* [*whatever*]. I'm like ... all these different people f**ked into one person ...' (*Enter the Ninja* 2009). Thus, it is not the Afrikaans working class that is the main theme of Die Antwoord's performance but South Africa itself: equally excluded from contemporary global culture, omitted and forgotten. What remains the most vivid example of this theme is the video of *Fatty Boom Boom* (2012), which offers a panoramic view on African cityscape. The tourist perspective adopted by the main character of the clip, namely a parodic representation of Lady Gaga, results in the ironic image captured in the video. While visiting Johannesburg, Africa's 'concrete jungle', Lady Gaga excitedly notices the 'most common' elements of the South African urban landscape: 'naughty hyenas eating rubbish', 'shop owner chilling with his black panther', and 'king of the concrete jungle, The Lion King' (2012). What also attracts her attention are 'some local musicians', namely Die Antwoord. 'Oh my God, look at their freaky fashion', exclaims Lady Gaga dressed in her raw-meat outfit (*Fatty Boom Boom* 2012). And indeed what a fashion this is; while Yo-landi's body is painted black and Ninja's body is painted in red and white, with what resembles an African ethnic war paint, DJ Hi-Tek, played by a black musician, wears a Ku Klux Klan suit, on which the words: 'kindness', 'joy', 'care', 'love', 'respect', 'faith', and 'peace', are written. The image of South Africa presented by Die Antwoord appears to be an empty phantasm filled with the stereotypes imposed by the Western world. The group exposes the popular culture as a neocolonial medium, a form of cultural dominance of the West. Die Antwoord seems to operate with the darkest fears of the Western world about their forgotten colony, serving an ironic answer, *Die Antwoord*, for the silent question about the postcolonial nature of the Western and non-Western relations (Figure 3.1).

The severe criticism of contemporary Western culture is expressed in a more direct fashion as well. Apart from the aforementioned pathetic

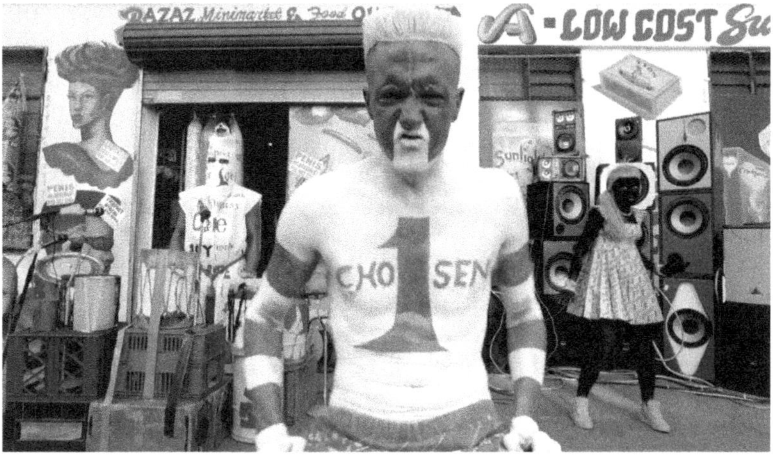

Figure 3.1 Die Antwoord as Johannesburg street musicians in the video *Fatty Boom Boom*

character of Lady Gaga, who explores concrete jungle in her meat dress and is eventually eaten by a lion, the *Fatty Boom Boom* video depicts also a piece of street art, which represents a nightmare that Ninja once had and reflects Die Antwoord's view on contemporary pop culture. The painting shows a 'Satan' hydra with the heads of Nicki Minaj, Kanye West, and five other mainstream rap stars. While Archbishop Desmond Tutu tries to exorcise the beast with a 'nine-millimetre Bible' (*'Fatty Boom Boom' – The Making of* 2012), the daemon defecates with the members of Black Eyed Peas. Desmond Tutu, the fierce opponent of apartheid, fights against popular culture as if it was a new form of cultural exclusion, repression of individuality and locality.

'What happened to all the cool rappers from back in the day?/ Nowadays all these rappers sound exactly the same', complains Ninja (*Fatty Boom Boom* 2012). The construction of authenticity is particularly significant for hip hop culture: one of the binary oppositions that rap operates with is the conflict between those who 'kept it real' and those who 'sold out' (Huq 2006: 110). That is why Ninja and Yo-landi dismiss two most renowned white rappers, Vanilla Ice and Eminem, in their defiant follow-up. 'Hi! My name is .../ Yo-Landi f**ken Visser/ Fight fight fight!' (*Fatty Boom Boom* 2012), cries Yo-landi in her direct reference to Eminem's 1998 song *My Name Is* and his erstwhile aggressive rap persona, while Ninja scorns Vanilla Ice's *Ice Ice Baby* lyrics: 'No, I do not want to stop, collaborate or listen' (*Fatty Boom Boom* 2012). Die Antwoord attacks mainstream in general (to recall again one of their songs 'Fok Julle Naaiers') by contrasting it with conceptual, refined art, which embraces locality.

References to white rappers, which appear in *Fatty Boom Boom*, address the problematic issue of whiteness and rap. Bennett implies 'the essential "blackness" of hip hop' (Bennett 2001: 93), while Omoniyi notices its intimate connection to a particular race (Omoniyi 2009: 115). Marx and Milton observe that 'for many black critics, the appropriation of the language and style of black-identified music forms and their attendant (sub)cultures by white musicians and audiences represents a form of cultural piracy' (Marx and Milton 2011: 740), which, according to Haupt, is due to 'the audience's recognition of the irony of a *white* rapper performing to the commercial's protagonist, a *black*, middle-class male partygoer, and his friends' (Haupt 2012b: 21). Hence, when white artists adopt a style and technique generally associated with black-identified music forms, like rap, the issue of cultural appropriation arises. Eminem or Vanilla Ice characters could be seen as modern minstrels or 'wiggas', namely 'the so-called white niggas who ape blackness by acting "hip hop" in dress, speech, body language, and, in some cases, even gang affiliation' (Tate 2003: 8). What seems to play an important role here are the rules of the music industry and the demands of the market, which led to commercialisation of the hip hop culture.

The problem that Die Antwoord exposes is even more complicated in regard to the racially ambiguous personas of its musicians: Ninja, being 'all these different people f**ked into one person' (*Enter the Ninja* 2009) and black faced Yo-landi (*Fatty Boom Boom* 2012). The ambiguity, which Ninja and Yo-landi promote, problematises racial identity and particularly its depiction in the popular imaginary. Transtextuality of hip hop culture and its inter-cultural coding is particularly noticeable in *Fatty Boom Boom*, which not only refers to Lady Gaga, Archbishop Desmond Tutu, or Vanilla Ice, but also to other pop cultural texts, such as the rap song of Masta Ace: 'Murder murder murder/ kill kill kill!' (*Slaughtahouse* 1993), or the movie *Die Hard* with the famous quote: 'Yippie-ki-yay motherf**ker!' (*Die Hard* 1988). 'The band's parodic intertextuality' described by Haupt (2012b: 124) results in the creation of 'an intentional hybrid', which, as Bakhtin noted, conceals an argument between two languages: the parodying language and the parodied one (Bakhtin 1981: 75). Through the recontextualisation of various cultural elements into one phenomenon, Die Antwoord represents a unique hybridisation between global and local influences. Representing simultaneously black and white, popular and niche, high and low, the project embraces ambiguity as well a inauthenticity, a characteristic trait of popular art. The aim of this strategy is to exaggerate the falseness of Western mainstream culture and expose its influence on South African consciousness.

Zef: parody of a parody?

It is more accurate to describe the music of Die Antwoord as *zef* style than hip hop itself. 'For us *zef* is, like, an ultimate style', a 'flavour of South Africa', claim Ninja and Yo-landi in the mock-documentary *Straight from the Horse's Piel* (2010). *Zef* with its ironic character that combines 'common' with 'cool', seems to be a hybridised form of hip hop, which resulted from merging the hip hop genre with local influences. The aim of relocating hip hop into *zef* territory is to express the marginalised, complex, and elusive identity of post-apartheid society:

> In the past you asked a person overseas what's the first word that comes to mind when thinking about South Africa and they go: 'apartheid'. But now, we are in f**king future now and when you are asking f**king people from the overseas, what's the f**king first word when they think of South Africa and you know what they will say? '*Zef*'. (*Straight from the Horse's Piel* 2010)

Hip hop and *zef* function as a frame for reinterpretation of identity and place, itself becoming a site of 'identities-in-translation' (Longinovic 2002: 7). The aim of this translation is to dismantle existing identities, which refer to post-apartheid order, and to propose alternative ones, based on dynamics of *zef*. What *zef* designates is a complex concept of parody and transformation that distances South Africans from their apartheid past. It connotes 'faux anger', and is yet 'comic and self-denigrating' (Krueger 2012: 401).

For some researchers, the discovery of the travestying character of Die Antwoord greatly complicates the problem of the representation of South African underclass (Haupt 2012a; Truscott 2011). With all its members having privileged backgrounds, their purported marginality seems rather misrepresented. The parodic element of Die Antwoord distances the group from the object of their representation. Because of the distance achieved, this form of misrepresentation seems different from that of other white rappers such as Vanilla Ice and Eminem. The identity

Die Antwoord performs is not limited to one class or race; it aims at the meta-identity of South Africa itself, which combines seemingly discordant elements: 'a lot of different things ... blacks, whites, coloureds, English, Afrikaans, Xhosa, Zulu – *watookal* [*whatever*]' (*Enter the Ninja* 2009). Die Antwoord's strategy questions their class position suggesting the possibility of going beyond their own class and race. They postulate abandoning class-related privileges in order to propose new vision of South African society as 'the rainbow nation' (*Faggot* 2012), where neither race nor class play a significant role. What I argue is that the notion of *zef* precisely defines Die Antwoord's performance, which cannot be reduced to the mockery of working-class white Afrikaners. What seems of utmost significance here is the self-parodying, self-negating aspect of *zef*. The meta-parody of *zef* mocks the mere concept of fixed identity imposed by social organisation. In its deconstructive attempts, *zef* targets the apartheid past, criticising it for its oppressive categorisation, the attempt to separate and evaluate. What is its prime objective however is opening of the present to the future: introduction of elements that are 'fresh', 'next level', and 'heavy futuristic'.

Truscott sees the travestying aspect of *zef* as a means to overcome melancholia, resulting from the 'loss of the apartheid' (Truscott 2011: 90). Following Freudian psychoanalysis (Freud 1923), Truscott argues that parody has a therapeutic value in overcoming the trauma of post-apartheid, since, for South African society, the 'loss of apartheid' meant also the loss of self-identity. Parody forms a strategy of distancing oneself to the parodied object. It challenges the past, proposing new ironic images, new understanding and evaluation of the past events without denying them. It is the attempt to create a new form of identity and, thus, it opens a possibility of transgression. Die Antwoord constantly refers to a past trauma, which restricts the future. *Zef* with its disruptive potential allows revaluating the past and focusing solely on what is to come, thus liberating South African society form the necessity to repeat.

'Is Die Antwoord a Blackface?', asks Haupt rhetorically (2012a). Haupt refers to Yo-landi's face painted black and the phenomenon of 'blackface minstrels', who for the sake of entertainment have imitated black culture operating mainly within racial stereotypes and colonial projections. 'Blackface minstrelsy, therefore, generated a range of meanings that could not be controlled by *black* subjects, given that they did not have the means to provide competing representations of *blackness*. Blackface revealed less about *black* subjects and more about *white* racist projections of *black* identities', explains Lott (1993: 101). The tradition of blackface has resulted in perceiving African-America's and Africans through the prism of invidious prejudice, analogously to the Orientalistic perception of the Middle East and North Africa examined by Said (1979). Hence, Haupt's main argument derives from the notion of cultural appropriation: Die Antwoord's 'mimicry' of mixed-raced identities produces a range of meanings or associations of black culture which cannot be controlled by the communities that the band caricatures in service of its artistic project (Haupt 2012b: 127). According to Haupt, Die Antwoord reinforces the tradition of false representation, imposing a white voice on the black subaltern community. As I have already stated, Die Antwoord exposes its own role of imposition on black music genre and the white underclass community. By the means of exaggeration, they uncover and criticise the mystification and falseness concerning the images promoted by popular culture.

Other arguments against treating Die Antwoord in terms of 'blackface minstrels' derive from the hybridised and transgressive nature of hip hop. It is not the race discourse that forms a core of hip hop culture but the 'race metaphor' (Simeziane 2010), which refers to the underclass rather than toa particular race itself. The race metaphor, being 'a salient reference point around which most hip hop artists seem to construct their identity' (Simeziane 2010: 116), indicates the process of relocation. Thus, Die Antwoord's racial references and exploration of taboo words, such as 'nigga' or 'faggot' (*Fok Julle Naaiers* 2012), a practice which emphasises a profound difference between United States and South Africa – 'the rainbow nation' (*Faggot* 2012) – where neither race nor language is negatively marked. Parodying certain language and certain imaginary, Die Antwoord deconstructs Western clichés concerning its lost colonies and criticises colonial relations preserved in contemporary culture. What Die Antwoord presents is the distorted image of Western culture that is reflected in the ironic and grotesque manner of Afrikaans style. The blackface of Yo-landi serves exactly this very purpose: it discusses and ridicules the tourist perspective on South Africa, represented here by Lady Gaga, indicating the impossibility of real African experience. Recalling numerous examples of visual abuses, Die Antwoord ironically plays with Western imaginary, deconstructing it entirely and reconstructing it on their own terms. The band challenges fixed identities imposed by Western mainstream culture in an attempt to redefine what it means to be South African. The answer they arrive at is only negative, for Africa escapes unambiguous definition and thus remains a dynamic phenomenon.

Haupt's argument exceeds the problem of misrepresentation of race. It is the position of social privilege that makes Die Antwoord's parody problematic, even for the researchers that appreciate their performance (Truscott 2011: 104). According to Haupt, what truly differs Die Antwoord's artists from the subject they seem to represent is 'their social-media savvy', and 'their access to high-quality media-production tools' (Haupt 2012b: 115). Their promotional potential is too high and the sophisticated aesthetic of Die Antwoord videos as well as the elaborate aspect of their performance do not fit their 'white-trash' personas and the quality of the lyrics. Haupt seems to claim that all those elements hamper the 'authenticity' of their endeavour, making representation of African working-class inadequate. As I have argued, it is not the identity of black South Africans or poor Afrikaners that Die Antwoord address in their performance. Their aim is to critique South Africa and its mixed identity by showing it as being in a constant state of flux.

Conclusions: freak show, or the answer

As I have argued, what Die Antwoord promotes is *zef*. The negative answer, *zef*, becomes replaced with the positive project: the uncertain identity of a freak, which resides beyond classification and power relations. It is identity of spectacle and ambiguity that enables rejection of existent identities and the reinvention of South African society. A positive aspect of Die Antwoord's performance is the construction of 'a carnivalesque sphere of liberation' (Truscott 2011: 102)', 'a liminal space', space in-between, where cultural influences intersect and hybrid, freak, identities are embraced.

The group offers a unique hybridisation between Western hip hop culture and local influences to produce a 'freak show', a complex entity of what is unfitting, unsettling, and out of place (du Preez 2011: 102). Die Antwoord seems to explore the notion of hybridity with their uncertain identities and monstrous bodies. In the video *Evil Boy* (2010), Ninja exposes his enormous horse-size penis and his prawn hand, which refers to the character Wikus van der Merwe from the movie *District 9* directed by Neill Blomkamp (2009). It suggests Ninja's animal monstrosity and indicates his problematic racial status (Figure 3.2).

Nederveen Pieterse claims, that 'hybridity problematises boundaries' (Nederveen Pieterse 2001: 220) while for Turner, liminal individuals or entities are 'neither here nor there; they are betwixt and between the positions assigned and arrayed by law, custom, convention, and ceremony' (Turner 1969: 95). Shildrick argues that 'the figure of the

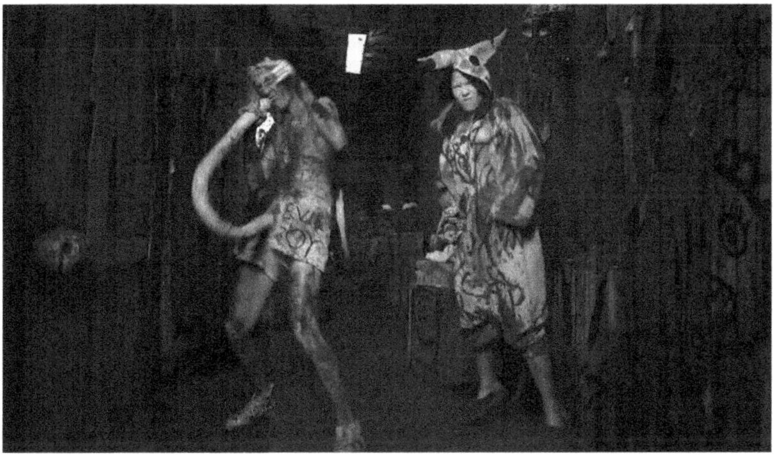

Figure 3.2 Ninja emphasises his creative potency by showing his enormous penis and using it as a microphone in the video *Evil Boy*

monster, like that of the freak, destabilises and problematises attempts to separate the category of the normal from the monstrous, the self from the other' (Shildrick 2002: 17). Die Antwoord's freak show destabilises the notion of identity based class or race, eroding the boundary between kitsch and art, between Western and African culture.

With the anthem of the excluded: 'I fink you freeky and I like you a lot' (*I Fink U Freeky* 2012), Die Antwoord provokes a discussion on the subaltern status of South African culture and uncertain status of South African identity. As music participates in and articulates identity discourse, it 'can be read as text for the manner in which it (en)codes, signifies, and/or textualises identity concerns and tensions' (Laubscher 2005: 308–311). The concerns debated in *I Fink U Freeky* regard the relation between the orientalising imaginary of the West and the complexity of African identity. The foreign, colonial perspective is clearly visible in quasi-ethnographic compositions; the video is overloaded with depictions of filth, revulsion, atrocity, ugliness, and deformation. The message it conveys seems to suggest that in the wake of the worst colonial nightmares, Africa returns the imperial gaze with anger. Alternatively, it could be argued the video tries to re-evaluate all the said elements. With the recurring verse: 'I fink you freeky and I like you a lot', Die Antwoord acknowledges the weird, dark and alien creatures that inhabit their videos. Hence the hybridity of the characters, of the

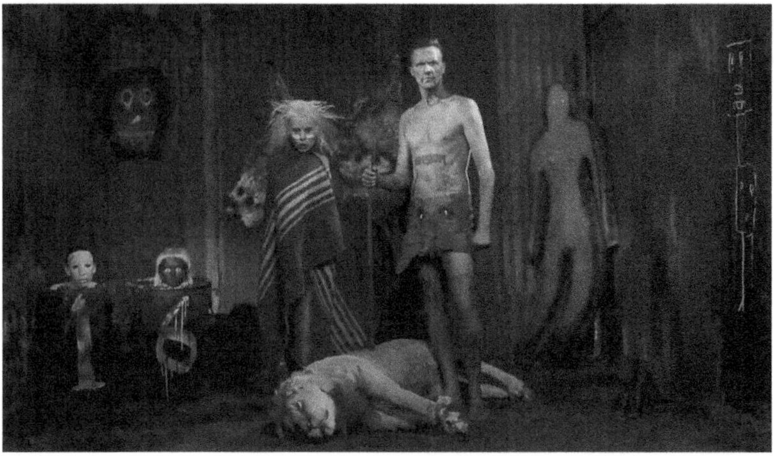

Figure 3.3 An frame from the video *I Fink U Freeky*

man with the bird head or the boy with the animal trunk, is recognised and embraced (Figure 3.3).

What constitutes the positive project of Die Antwoord is the pure heterogeneity and hybridity that freaks connote. Through the hybridisation of the *zef* genre and the use of 'race metaphor' (Simeziane 2010: 116) Die Antwoord aims at transforming South African identity as an 'identity-in-translation' (Longinovic 2002: 7) with the constellation of its elements being constantly reconfigured. While Krueger argues that 'Die Antwoord seems ironically not to provide any answers at all' (Krueger 2012: 407), and du Preez asks: 'With regard to Die Antwoord, which speaks from bastard threshold, what post-hegemonic potential do they bear?' (du Preez 2011: 114), I claim that Die Antwoord uses hip hop means of mixing and sampling to re-imagine South African aesthetics, reconstruct post-apartheid identities and eventually, embrace a freak identity that escapes defined categories of class and race.

Acknowledgement

This work was supported by the Foundation for Polish Science – International PhD Programme, co-financed by the European Union within the European Regional Development Fund.

Notes

1. Die Antwoord's live performances, especially the early ones, have been often accompanied by black-and-white visualisations. Contrarily to the contrasting sophistication of their musical videos, raw character of said visualisation supports the low quality of their rap.
2. Although Die Antwoord's phenomenon comprises both elements of hip hop and rave, I limit my analysis of their music to the hip hop element. I follow Huq in claiming that rave signifies mainly 'the politics of pleasure' (Huq 2002: 93) and escapes any other political organization. It is impossible to decide whether the communities created through decentralisation of individual identity in dance trance have political agency, and whether the repetition of rave beats has an affirmative character.

Works cited

Androutsopoulos, Jannis (2009). 'Language and the three spheres of hip hop', in H. Samy Alim, Awad Ibrahim, Alastair Pennycook (eds.), *Global Linguistic Flows. Hip Hop Cultures, Youth Identities and the Politics of Language* (London and New York: Routledge), pp. 43–62.

Bakhtin, Mikhail (1981). *The Dialogic Imagination: Four Essays*. Translated from the Russian by Caryl Emerson and Michael Holquist (Austin: University of Texas Press).

Ballen, Roger (2005). Shadow Chamber (London: Phaidon).

Ballen, Roger (2014). Asylum of the Birds (London: Thames & Hudson).

Bennett, Andy (2001). *Cultures of Popular Music* (New York: Open University Press).

Bhabha, Homi K. (1994). *The Location of Culture* (London and New York: Routledge).

Connell, John and Chris Gibson (2001). *Sound Tracks. Popular Music, Identity and Place* (London and New York: Routledge).

Die Antwoord (2013). *About* [Online]. www.dieantwoord.com/tension.html#about (accessed 5 April 2013).

Forman, Murray (2004). ' "Represent:" Race, space, and place in rap music', in Murray Forman, Mark Anthony Neal (eds.), *That's The Joint! The Hip-Hop Studies Reader* (London and New York: Routledge), pp. 201–222.

Freud, Sigmund (1923 [1953]). 'The ego and the Id', in *The Standard Edition of the Complete Psychological Works of Sigmund Freud*, edited and translated by James Strachey, vol. 19 (London: Hogarth Press), pp. 12–66.

Freud, Sigmund (1919 [1953]). 'The uncanny', in *The Standard Edition of the Complete Psychological Works of Sigmund Freud*, edited and translated by James Strachey, vol. 17 (London: Hogarth Press), pp. 219–252.

Haupt, Adam (2012a). 'Is die Antwoord blackface?', *Safundi: The Journal of South African and American Studies* 13, pp. 417–423.

Haupt, Adam (2012b). *Static. Race and Representation in Post-Apartheid Music, Media and Film* (Cape Town: HSRC Press).

Huq, Rupa (2002). 'Raving, not drowning: Authenticity, pleasure and politics in the electronic dance music scene', in David Hesmondhalgh and Keith Negus (eds.), *Popular Music Studies* (London: Arnold Publishers), pp. 90–102.

Huq, Rupa (2006). *Beyond Subculture. Pop, Youth and Identity in a Postcolonial World* (New York: Routledge).
Krueger, Anton (2012). 'Zef/poor white kitsch chique: Die Antwoord's comedy ofdDegradation', *Safundi: The Journal of South African and American Studies* 13, pp. 399–408.
Laubscher, Leswin (2005). 'Afrikaner identity and the music of Johannes Kerkorrel', *South African Journal of Psychology* 35, pp. 308–330.
Lipsitz, George (1994). *Dangerous Crossroads. Popular Music, Postmodernism and the Poetics of Place* (London and New York: Verso).
Longinovic, Tomislaw Z. (2002). 'Fearful asymmetries: A manifesto of cultural translation', *The Journal of the Midwest Modern Language Association* 35, pp. 5–12.
Lott, Eric. (1993). *Love and Theft: Blackface Minstrelsy and the American Working Class* (New York: Oxford University Press).
Marx, Hannelie and Viola Candice Milton (2011). 'Bastardised whiteness: "Zef" -culture, Die Antwoord and the reconfiguration of contemporary Afrikaans identities', *Social Identities: Journal for the Study of Race, Nation and Culture* 17, pp. 723–745.
Murray, Conrad Derek. 'Hip-hop vs. high art: Notes on race as spectacle', *Art Journal*, 63, pp. 4–19.
Nederveen Pieterse, Jan (2001). 'Hybridity, so what? The anti-hybridity backlash and the riddles of recognition', *Theory, Culture & Society* 18, pp. 219–245.
Omoniyi, Tope (2009). ' "So I choose to Di Am Naija style." Hip hop, language, and postcolonial identities', in H. Samy Alim, Awad Ibrahim, Alastair Pennycook (eds.), *Global Linguistic Flows. Hip Hop Cultures, Youth Identities and the Politics of Language* (London and New York: Routledge), pp. 113–138.
du Preez, Amanda (2011). '*Die Antwoord Gooi Zef.* Liminality: Of monsters, carnivals and affects', *Image & Text* 17, pp. 102–118.
Rose, Tricia (1994). *Black Noise: Rap Music and Black Culture in Contemporary America* (Hanover and London: Wesleyan University Press).
Said, Edward W. (1979). *Orientalism* (New York: Vintage Books).
Simeziane, Sarah. (2010), 'Roma rap and the black train: Minority voices in Hungarian hip hop', in Marina Terkourafi (ed.), *The Languages of Global Hip Hop* (London, New York; Continuum), pp. 96–119.
Shildrick, Margrit (2002). *Embodying the Monster: Encounters with the Vulnerable Self.* (London: Sage).
Straight from the Horse's Piel (2010). Directed by Kobus Holnaaier [Film]. vimeo.com/15196261 (accessed 27 August 2014).
Take No Prisoners (2010). Added by Die Antwoord [YouTube video]. www.youtube.com/watch?v=vx1cYUb-0f4 (accessed 27 August 2014).
Tate, Greg (2003). 'Nigs R us, or how blackfolk became fetish objects', in G. Tate (ed.), *Everything But the Burden: What White People are Taking from Black Culture* (New York: Broadway Books), pp. 1–14.
Terkourafi, Marina. (2010). 'Introduction: A fresh look at some old questions', in Marina Terkourafi (ed.), *The Languages of Global Hip Hop* (London, New York: Continuum), pp. 1–18.
Thompson, Krista (2009). 'The sound of light: Reflections on art history in the visual culture of hip-hop', *The Art Bulletin* 91, pp. 481–505.

Truscott, Ross (2011). 'National melancholia and Afrikaner self-parody in post-apartheid South Africa', *Psychoanalysis, Culture & Society* 16, pp. 90–106.
Turner, Victor (1969). 'Liminality and communitas', in *The Ritual Process: Structure and Anti-Structure* (Chicago: Aldine Publishing), pp. 94–130.
Zef Side (2009). Directed by Sean Metelerkamp [Film]. www.youtube.com/watch?v=Q77YBmtd2Rw&index=16&list=FL6DicDItkLQ7N4ag1JS5mqg (accessed 27 August 2014).

4

Recycled Music for Banal Nation: The Case of Serbia 1999–2010

Srđan Atanasovski

In this chapter I address the ways in which popular music genres have been recycled in the Serbian post-socialist political landscape. Specifically, I analyse how Western-styled music production was relocated, both spatially and temporally, from being a vehicle of purported 'freedom of expression' in socialist Yugoslavia, to operating as a mechanism of Serbian banal nationalism. During the 1990s, Serbian nationalism emerged as an antagonistic force playing a crucial and dominant role in the violent breakup of Yugoslavia and aiming to retain control of vast swathes of land which were (in Serbian nationalistic discourse) perceived as parts of Serbian national territory. The aggressiveness of Serbian nationalism was reflected in various popular music genres, not least in the infamous turbo-folk. Thus, it is important to trace the mechanisms whereby certain products of popular music, through processes of spatiotemporal relocation, were employed in order to banalise the 'hot' Serbian nationalism, and represent it in a different, Westernised light. I will show how the pop and rock music became engrafted into seemingly innocuous representations of Serbian patriotism, or 'civic nationalism', through widely accepted practices which were even perceived as 'above the political' in the everyday jargon.

Michael Billig (1995) defined banal nationalism as nationalism which is reproduced in everyday, seemingly banal and innocuous situations, thus becoming the omnipresent doxa and discourse that we use in order to make sense of our everyday lives. Through its banality, nationalism is transformed into 'patriotism', stripped of its negative connotations and recognised as something positive. However, Billig positions his analysis of banal nationalism in contrast to what he terms 'hot nationalism' – nationalism which is violent, out of control and, importantly, which happens outside the Western, civilised world.

Between banal and hot nationalism there is a clear colonial link: nationalism, in its 'positive', banal form, appears as something which is inherently Western, produced and originally practised in the civilised realm of Western countries, whereas hot, violent nationalism thrives on the edges of our civilisation. Drawing on this perspective of Billig's research,[1] I will explore how state apparatus deals with situations of thriving 'hot nationalism'. I argue that banalisation is a dynamic process, a mechanism through which the state strives to reaffirm its own political authority and reducing nationalism to what is seen as a mere demonstration of patriotism. In this regard, the banalisation of nationalism is also a colonial phenomenon, a process through which the societies which perceive themselves as the colonial 'Other' struggle to emulate the political models of the Western world.

I take example of post-Dayton Serbia[2] and two state campaigns which were conducted between 1999 and 2010: the first one concerns organising popular music events in public spaces during the NATO bombing in 1999, and the other is the media and television campaign of rebuilding the Avala TV Tower, destroyed in the NATO bombing of 2005. Both these situations can be described as 'hot nationalism'. In the first example, during the final phase of Milošević's regime in Serbia, the country was ostracised by most of the international community, while its administration was involved into forcible displacement and other ethnically inspired crimes in the territory of its former autonomous province of Kosovo.[3] In the second case, passionate nationalistic discourse was not only present through dealing with the traumatic legacy of NATO bombing, but also in connection with the open issue of Kosovo's secession, which was interpreted as an onslaught on the Serbian nation *in toto*.

In my analysis I look into the ways that the nationalistic project and nation-state intertwine. Modern nationalism, based on the doctrine of sovereignty and the idea of statehood is deeply ingrained in every political project presenting itself as national (Yack 2012: 136). However, practices of nationalism are continuously being formed and exercised in-between state and private realms, and 'national identity' incorporates both a sense of belonging to the nation and a sense of loyalty to the nation-state. Through my examples of the use of Western-styled popular music genres in state campaigns in Serbia, I show how state apparatus finds ways to appropriate sentiments provoked by the 'hot nationalism' (produced in the private realm of 'national belonging') through such mechanisms as banalisation, to solidify the attachment of the individuals to the nation-state.

In the first part of this chapter I review discourses of nationalism in Serbian popular music in both Yugoslav and post-Yugoslav contexts. I also consider music's positioning within Yugoslav and Serbian state apparatus, in particular to point out purported 'Eastern' and 'Western' allegiances of certain genres of Serbian popular music. This is of importance as most of the music which was used in the campaigns analysed was originally produced in a different context, and continues to carry existing meanings and connotations. I will then analyse two named campaigns showing how mechanisms of banalisation were employed by recycling existing music. In my conclusion I will point to similarities between these projects, notwithstanding their ostensibly different political contexts.

Serbian popular music between the East and the West

Both of the projects that I examine resorted to 'recycled music' and it is important to understand how the decisions what to recycle were made and the symbolic legacy of the music which became part of these campaigns. Serbian national identity has been riven with different interpretations of notions of 'West' and 'East', a dichotomy constantly used in the labelling of popular music genres. These labels, part of the mechanisms of manipulating identities, were used as mechanisms of governing, as a technology of power. From its inception in the early nineteenth century, the intellectual elite defined Serbia's identity as non-Ottoman/non-Eastern, in an effort to vindicate its territorial enlargement at the expense of the empire in decline. This was especially visible in the attitudes of various composers and ethnomusicologists who strove to purify Serbian music by eliminating what they perceived as Ottoman elements. However, in the context of Yugoslavia, throughout the twentieth-century, Serbian identity was also defined as non-Western, which in this instance was defined as 'traditional', pro-Russian, Byzantine, and Orthodox Christian. This became especially obvious in the 1990s, during the breakup of Yugoslavia, which was portrayed as the 'dungeon of the Serbian nation', and Western powers were seen to be allying with the other sides in the conflict (Croats, Bosniaks, Kosovars, etc.).

The creation of socialist Yugoslavia after the Second World War entailed a significant curbing of the particular nationalisms of 'Yugoslav nations', including Serbian, and their subordination to Yugoslav patriotism which was reformulated through the Communist Part of Yugoslavia's ideology of 'brotherhood and unity' (Calic 2010: 180–183). After the break with the Soviet Union in 1948, Yugoslavia increasingly tried to portray itself as a state where socialism had a more human face and one more open to the influences from the West, including its popular music, which was by far more captivating than the Soviet-modelled mass songs (Doknić 2013: 271–279; Marković 1996; Vučetić 2006, 2012). Thus, while central government tried to curb perceived outbursts of particular nationalism within Yugoslavia, it fostered Western-influenced popular music, first in the form of jazz and Italian pop and afterwards global rock and roll. Already in 1948 the Communist Party's attitude towards jazz (which was ostracised throughout the Soviet political sphere of influence) became increasingly lenient, and in 1954 US diplomats noted that 'surprisingly large space is awarded to American jazz music' on the Yugoslav radio programmes (Marković 1996: 475, Vučetić 2006: 174–176). During the 1950s and 1960s a vast numbers of songs performed by Yugoslav singers were translated and adapted from Italian. Examples include *Nogometna utakmica* by Beti Jurković, released by the Zagreb production company Jugoton in 1963, just a year after the original Italian version *La partita di pallone* (composed by Edoardo Vianello and performed by the celebrated Rita Pavone). Songs in similar fashion were widely arranged by Yugoslav composers and made popular through a system of pop-music festivals modelled on Sanremo, including ones in Opatija, Split, Belgrade, and Sarajevo. From the mid-1960s, rock and roll started to dominate the Yugoslav music scene, replacing both jazz and Sanremo festival-inspired popular music, although it was sometimes perceived as subverting the socialist regime, primarily because its promotion of sexual freedom went against the regime's strict moral values, Yugoslav rock and roll was benefiting from state financial and logistical support throughout the 1970s and 1980s. YU grupa presents a paradigmatic example, as one of the pioneering Yugoslav rock bands, formed in the autumn of 1969 (Ivačković 2013: 162). Although promoting a music genre perceived as Western, YU grupa integrated into the ideological apparatus of the socialist Yugoslavia (for example, holding annual celebration concerts for the Republic Day on 29 of November, symbolically the most important holiday of the new regime) and by combining rock and folk music elements. The latter became known as 'shepherds' rock' ('pastirski rok'), or 'folklore YU rock' (Đorđević 2002: 120) and was largely responsible for the wide popularity of rock music in Yugoslavia, seemingly cutting through the class barriers. It helped the Yugoslav state create the unique image of a socialist state open to foreign influences and participation in global popular and consumer culture and one which awards it citizens with a latitude of 'freedom of expression'. Although *de facto* part of the state system, rock musicians still relied on semi-staged scandals, playing with the social taboos and even provoking some local censorship incidents (Arnautović 2012: 161). In this way they were able to present themselves as the harbingers of new freedom in Yugoslav society. The rock culture which developed in Yugoslavia can in general be regarded as an example of 'colonial mimicry' (Bhabha 1994: 85), where this society, which stood on the edge of the socialist realm of the Eastern Europe, tried to participate in what was perceived as the free and civilized world of the West, producing music which was 'almost the same, but not quite'.

At the same time these, Western-styled popular music genres had to compete in the market with so-called 'newly composed folk music' (Rasmussen 2012). This was inspired by folk music (primarily of Serbia and Bosnia), but the genre also relied on newly composed tunes, arrangements drawing on pop-music models, as well as on 'stardom effect' (Gledhill 1991) and the huge popularity of the leading performers. Primarily characteristic of the Serbian scene where it quickly became the most profitable music genre (Arnautović 2012: 183), it was repeatedly decried in public discourse, both by intellectuals and party officials as backward and 'Eastern'. Moreover, it was associated with the extensive migration from rural to urban areas, as part of the great socialist project of industrialisation and hence was subjected to rampant stigmatisation and discrimination (Kos 1972). Some of the best-selling singles in former Yugoslavia's popular music history were created by performers such as Tomislav (Toma) Zdravković, who built his reputation on being a true melancholic bohemian of Serbian kafanas, and Zilha Bajraktarević (known as Silvana Armenulić), who shocked the public with highly sexualised appearances in the media. The 1980s saw a significant wave of Serbian nationalistic revivalism, as the intellectual elite was frustrated by the implementation of the 1974 constitution. This gave enhanced rights to the autonomous provinces within Serbia, Vojvodina and Kosovo (officially 'Kosovo i Metohija'), the latter populated by the Albanian majority who supposedly started terrorising the Serbian minority (Sundhaussen 2007: 379). Metanarratives of Serbian greatness such as the Kosovo myth, were resurrected via reference to medieval Serbian history and First World War narratives, where the Serbian nation was victimised, decimated, but finally victorious in combat. Unlike Yugoslav rock and roll, the newly composed folk music embraced this nationalistic turn. This was achieved by incorporating First World War regalia into its visual representations and by drawing on certain songs thematically connected to these national narratives. A striking example was Predrag Živković

Tozovac's single *Jeremija* (1972), which controversially draw on narratives of a victorious Serbian army in the First World War. Tozovac posed on the sleeve in the pre-war uniform of the Kingdom of Serbia holding a rifle, and for the reissue, in a semi-comic twist he even mounted a heavy artillery gun. Some of the most poignant examples are songs and compilations produced in connection with the six hundredth anniversary of the battle of Kosovo Field, in 1989, which opportunely coincided with the new rise of Serbian nationalism, for which the narrative of this medieval battle has served as one of the cornerstone myths. The celebration again strengthened this link between newly composed folk music and Serbian nationalism. This is primarily reflected in the immense popularity of the song *Vidovdan* ('St. Vitus' Day'), performed by Gordana Lazarević and composed by Milutin Popović Zahar, which became constant fixture of every social gathering, ranging from political rally to wedding celebrations. The poignant verses of the song create a sense of timelessness and speak of the protagonist's primordial connection with the land of Kosovo, as a core part of the Serbian national territory ('As the eternal flame in our hearts / The truth of the Kosovo Battle remains / Wherever I go I come back to you / Who could steal Kosovo from my soul'). Thus, newly composed folk music, as the music genre which supposedly resisted the 'Westernisation' of Yugoslavia, simultaneously evolved into the vehicle of increasingly 'hot' Serbian nationalism.

With the escalation of violence in Yugoslav territories during the early 1990s, and increasingly belligerent Serbian nationalism, perception of Serbia as belonging to the 'Eastern' realm was particularly accentuated. In the transformations which took place following a deepening economic crisis, the production of folk music was handed over to newly founded private market-orientated TV and production companies. The Serbian popular music scene witnessed an aggressive burgeoning of 'turbo-folk' music, a genre emanating from newly composed folk music. Producers openly pursued popularity and profit by drawing on music patterns found in the popular genres across the Balkans and the Near East and ones characteristic of Western dance-pop, as well as sex-and-violence and glittering media representations. Songs by dance-pop performer Ivan Gavrilović are often regarded as the aesthetic manifestos of the new genre: *200 na sat* ('200 km per hour'), *Motori* ('Motorcycles') and *Hoću s tobom da đuskam* ('I want to dance with you') extol dangerous and sexually charged night-life behaviour. Importantly, turbo-folk embraced female star singers launching their careers as singers of newly composed folk music and introducing a genre of singing perceived as intrinsically folkloric and 'Eastern'. An example of this new blend is the song *Crveno* ('Red') performed by Svetlana Ceca Ražnatović, where the singer pairs her well-known singing style with its harsh timbre and wide vibrato with verses alluding to her precarious life style ('Go with me through the red light / What is hers will be mine / Tie your eyes and do not brake'). The song video accompanying the song also featured a highly stylised sexual performance.

The lack of the public support for the music industry had a devastating effect on rock music production, which was unsustainable in the fragile market. As the musicians could not secure their incomes in the new economic system, the market share of the genres which were perceived as the Western-influenced dwindled. Turbo-folk was vehemently deprecated by intellectuals, on similar grounds as newly composed folk music, with the strong accentuation 'Eastern' sound, often referred to as the 'Tehranisation' of Serbian (folk) music (Atanasovski 2012; Kupres 2004). A number of intellectuals and musicians, correlated the rise of turbo-folk with the nationalistic ethos of the belligerent Milošević regime. Certain musicians associated with this genre (such as Mirko Pajčin, known as 'Baja Mali Knindža') openly engaged in the conflict, advocating the Serbian cause through blatant nationalistic lyrics. Pajčin openly sympathised with Serbian paramilitary commanders in the wars in Croatia and Bosnia, expressing his nationalist and xenophobic attitudes towards Croats and Bosniaks. Moreover he supported the far-right Serbian Radical Party, whose political programme based on the irredentist project of so-called 'Greater Serbia', glorified its controversial leader Vojislav Šešelj. Alhough Pajčin's stereotypically folkloric music style may seem prosaic, the appeal of his songs is mostly produced by poignant verses which speak of Serbian racial supremacy and historical victimhood ('Ever since God created humans / always others come to judge us / always someone bothers the Serbs'), often directly insulting or threatening members of other national communities (Longinović 2001: 639). Svetlana Ceca Ražnatović herself became representative of these connections when she married Željko Ražnatović Arkan, one of the main leaders of Serbian paramilitary troops active in the war in Croatia and Bosnia. Some scholars argued that Milošević's regime itself produced or supported the rise of turbo-folk in order to destroy the cultural alternatives to the Serbian nationalistic project (Gordy 1999; Kronja 2000). However, no proof of explicit state involvement in the rise of turbo-folk can be found, and the process can be explained solely through the effects of economic crises and the imposition of the market economy (Đurković 2001). Thus, the loss of cultural alternatives described by Gordy was a product of major social and economic crises which mostly affected Western-influenced music genres, incapable of sustaining their activities without the governmental support they enjoyed in socialist Yugoslavia. I would argue that the radical nationalistic examples of turbo-folk, although representative of the ideological system which imposed itself as dominant in contemporary Serbian society, cannot be regarded as a state project, but more as instances of 'hot nationalism' operating outside the state apparatus.

'The Song Has Kept Us': public concert during the NATO bombing

The NATO Bombing of Yugoslavia began on 24 of March 1999, just a month after unsuccessful negotiations held in Rambouillet, where the government of Federal Republic of Yugoslavia[4] could not agree with representatives of the Kosovo Albanians and the international community, as it did not accept the NATO presence in Kosovo or the possibility of holding an independence referendum after three years. NATO unilaterally decided to bomb Yugoslavia in order to curtail its military and policing capacities. While the campaign wrought havoc on Kosovo, both because of intensive NATO bombardment and because of intensified ethnic cleansing perpetrated by Serbian forces, the mood in the capital of Belgrade was quite different. The citizens' daily lives were disrupted with frequent emergency notices and schools were closed. A major source of traumatisation was the sonic experience of the campaign. In the transfigured *soundscape* of the city relentless sirens announced imminent danger followed by the sound of heavy bombardment of military facilities and headquarters, political and government buildings in the city centre and on its outskirts. [5] The NATO forces would occasionally employ so-called 'sound bombs', a sonic effect of high-volume and deep-frequency caused by low-flying jets breaking the sound barrier. Responding to this *sonic terror* (Goodman 2010) the government deployed strategies to mobilise the citizens of Belgrade which were equally centred on the sound and sonic experience. Namely, the government organised public concerts of popular music in the open spaces in most of the cities of Serbia proper. The first concert was organised by Belgrade city council, on Sunday 28 March, just four days after the bombing started. Surprisingly, it was concentrated solely on rock music, featuring the groups Električni orgazam, Deca loših muzičara, Bajaga i Instruktori and YU grupa, as well as musicians such as Dejan Cukić and Rambo Amadeus. The musical style of these groups ranged from pop/rock to funk and (post-) punk rock, most of which had gained popularity in socialist Yugoslavia. Being quite immune to the rise of Serbian nationalism, these musicians were often connected to the common cultural identity of former Yugoslavia which permeated its rock scene. Previously mentioned YU grupa had the longest career so far, and the one most clearly connected with the former Yugoslav identity. The second generation of Yugoslav rock was characterised by a mollified sound, often aligned with seemingly depoliticised pop music. Such were Bajaga i Instruktori, whose popularity was particularly urban and Belgrade-based, as well as relying on the charisma of their lead singer, Momčilo Bajagić Bajaga, who was particularly masterful in interpreting emotionally charged rock ballads (Ivačković 2013: 328–337). However, among the performers present at the 28 March concert, Električni orgazam was also a prime representative of the third generation of Yugoslav rock, which accrued their popularity in the late 1980s. Renouncing any folk conceptions of 'shepherds' rock', they built their image on an aggressive rock sound and often politically engaged verses criticising contemporary Yugoslav society. Openly dissident, they became part of mainstream Yugoslav rock, illustrating the purported openness of this state-socialist society (Arnautović 2012: 163). Thus, the 28 March concert in Belgrade can be interpreted as an event where the full kaleidoscope of Yugoslav rock was relocated in its time, space, and meaning: the songs performed were mostly at least a decade old, and originally representative of the pro-Western values of Yugoslav socialist society. Here they were performed in an entirely new context, in a novel state whose governing apparatus reckoned with their accrued popularity and utilised them to banalise a highly charged political situation. This concert, which proved to be only the first in a series of daily organised events, was especially memorable because of to intense affective atmosphere in which musicians played despite air raid warnings. As a truly surprising event, the concert articulated the feelings of fear of both performers and the audience at the gathering thereby transforming them into a form of therapeutic defiance (Grujičić 1999).

The government pushed forward the idea of organising public concerts throughout the NATO campaign on a daily basis. Besides the daytime concerts organised in the centre of Belgrade by the city council under the slogan *The Song Has Kept Us*, the two ruling parties staged night-time concerts on two important Belgrade bridges, purportedly to serve as a living shield dispelling the danger of their destruction. The exclusively rock music character of the concerts did not last, but neither was a certain musical selection maintained. As the programme slid into musical mainstream and mishmash, including turbo-folk performers, most of the musicians invited by the government to perform felt forced to comply (Mira Škorić statement in Kupres 2004). While public support for the initiative started to dwindle, which was reflected in the numbers of visitors which seemed to wane from day to day, the organisers tried to gather as large an audience as possible by bringing pop rock stars Zdravko Čolić and Đorđe Balašević to play on the 11 April, Easter day. Somewhat similar to Momčilo Bajagić, Zdravko Čolić and Đorđe Balašević rose to great popularity in 1970s and 1980s respectively, positioning themselves as seemingly apolitical figures and appealing to a wide audience. The emblematic songs which were repeated through these concerts were published on a CD featuring the target logo that became an emblem of this campaign of supposed popular resistance. Published by anonymous ad hoc 'Target rekords', the CD was widely sold on the streets during the concerts and reflected their repertoire. Again, one notes the dominant presence of former Yugoslavian rock music, either left unchanged or adapted for this special purpose. The meaning of old Yugoslav rock, which was perceived as cosmopolitan, was here relocated into processes of banalisation of the Serbian nationalistic cause. The songs featuring by then the futile messages of Yugoslav identity, as the one built on youthful energy and cosmopolitan values (which was very much in dispute with the turbo-folk genre), such as *Igra roken rol cela Jugoslavija* ('The whole of Yugoslavia is dancing to rock and roll music') played by the Električni orgazam, dominate the compilation, which attracted their wide popularity. Some songs accrued different meaning in a changed context: the song *Avionu slomiću ti krila* ('Airplane, I'll break your wings') by former Yugoslav rock band Riblja čorba originally features a love theme (in which the 'breaking of the wings' of the passenger jet was used to express the protagonist's desire to be with his loved one). However the song's meaning was reinscribed via allusions to the shooting down of enemy planes, regularly reported in the media and widely celebrated. A special place was awarded to the song *Ringišpil* ('Carousel') by Đorđe Balašević, who wrote new war-themed lyrics to this popular love song from 1991. The original song, written in a formulaic and dreary singer-songwriter style, was structured as a call to a former lover to reappear and 'add some colour'. Balašević's interventions were mostly concentrated on the opening two stanzas, which set the melancholic mood in which 'it's all the same' to the protagonist. Indeed, Balašević uses the song's exaggerated melancholia to depict the imminent danger of the war as a mere nuisance and comparable to rain, thus using a love ballad as an instrument to dilute the seriousness of the precarious political situation. Specially produced songs with banal, patriotic verses were also added to this compilation, such as *Ja volim svoju zemlju* ('I love my country'), produced by City Records and featuring an array of pop and folk singers, and *Samo sloga Srbina spasava* ('Only unity saves the Serbs'), featuring Riblja čorba and rock musicians Bajaga and Dejan Cukić. It is evident that this repertoire conforms to the idea of 'banal nationalism'. Moreover, its banality is seen as kind of therapeutic vehicle which transforms the trauma of sound and fear provoked by the NATO bombing into complacency and stupor.

Đorđe Balašević, *Ringišpil*, first two stanzas, original version (1991) and war-time version (1999)

It's raining this morning since four or five, the sky decided to flood the world over the city for days more of the same scenery.	They're beating this morning since four or five, NATO decided to change the world over the city for days fly of the same rockets.
Pouring rain, but that's his thing, hey, it's all same to me, flat like the north of the Banat, more or less, both with and without rain.	Pouring NATO, but that's his thing hey, it's all same to me flat like the north of the Banat, there'll be war, so be it, we end up same.

'Let's Raise the Tower at the Avala'

After the fall of Milošević's regime in 2000, Serbian citizens were strongly divided between the poles of the nationalistic and pro-Western/pro-European cause, with the rift between generations being particularly pronounced. One could argue that the rift between supporters of Milošević's regime on the one hand, and its opponents on the other, was so wide that the integrity of the nation was brought to question. Members of these two poles defined themselves and allied to specific causes not only by using specific political rhetoric, but also by using patterns of cultural consumption to identify themselves. In the late 1990s, rock music was clearly associated with resistance against Milošević's regime, with examples of outspoken criticism present in the works of some of the rock musicians (Mijatović 2008). Music also occupied a prominent place in the anti-Milošević street protests, where some rock musicians and performers took part in the rallies and recorded songs with political and satirical verses. From the point of view of members of Milošević's opposition, it was regarded as natural and expected that rock musicians, who were pro-Western cosmopolites, unlike the turbo-folk performers, were among those who opposed Milošević's regime and the nationalistic and aggressive militant culture it represented. The ethos of rock music, as original, individual, and, last but not the least, pro-Western, was opposed to folk music on a political as well as on an aesthetic level. Folk singers purportedly continued with the 'Tehranisation' of Serbian music and, as individuals, they were either seemingly apolitical, connected to the ruling parties of the Milošević regime (such as Zorica Brunclik, prominent member of Yugoslav Left), or involved with famous nationalistic figures (such as Svetlana Ceca Ražnatović). The rock musicians, such as Rambo Amadeus, Darkwood Dub, and Eyesburn (Mijatović 2008), not only gave their support to the anti-Milošević movement, but also used their music aesthetic to back the protests. Particularly prominent was Dragoljub Đuričić, reputedly one of the leading rock drummers of former Yugoslavia (formerly working with YU grupa, Leb i Sol, Kerber, and as a member Zdravko Čolić's and Đorđe Balašević's bands), who embarked with projects in progressive rock based solely on percussion instruments. Playing portable drums in a percussion ensemble with his associates, Đuričić gained recognition by actually leading the street protest marches.

I would argue that the purpose of the project of rebuilding the Avala TV Tower was to answer the challenges of homogenising a divided nation by restoring the authority of the state. It also helped to banalise the new wave of Serbian nationalism stemming from dissatisfaction with the Kosovo secession, which led to the fall of Milošević. In this process, relocating the rock music – and more so, the musicians themselves – from an anti-Milošević position to seemingly legitimising the rejuvenated Serbian state – proved to be one of the crucial mechanisms. The Avala TV Tower was originally constructed between 1961 and 1965 and was the tallest structure in the Balkans, located on the outskirts of Belgrade. It was destroyed on 29 April 1999 during the NATO bombardment of Serbia. The campaign to re-erect the tower, which commenced in 2004, was permeated with the ethos of repulsion and repudiation towards the NATO bombardment, and the very act of rebuilding was to symbolise an annulment of the consequences of the action, the most serious being the secession of Kosovo. Importantly, the campaign was launched following violent unrest in Kosovo in March 2004, when many Serbian landmarks and buildings were torn down and part of Serbian population left the province, contributing to the anxiety of the remaining population and to the overall discontent of nationalists in Belgrade.

Although heavy financial and logistical support came from the state government, the funding mechanisms of the project tended to produce a semblance of private or non-governmental initiative. Radio Television of Serbia broadcast a series of live TV shows featuring SMS donations from the audience, and citizens were invited to help the project, fostering a feeling of solidarity towards a 'people's' cause. The restoration of the Avala TV Tower was imbued with emotional significance for many Belgrade citizens who were used to seeing this prominent structure in their everyday life as part of the healing process after the traumatic experience of NATO bombing. The reconstruction of Avala TV Tower is significant as it claimed to symbolically overwrite the effects and memory of NATO bombing, thereby restoring the unity of the state territory and reconstituting Serbian 'homeland' as an organic whole. This strategy of erasing history was prominent in the city mayor's opening speech, where he noted that it is was as if the tower was shrouded in mist for all these years, and only now could be seen clearly again. The tower symbolically and physically represented the organic unity of Serbian national territory, as (almost) all the parts can be thence seen in clear weather, as emphasised in the speech of one of the members of the cabinet on the same occasion.

Musical products incorporated into the project included a prominent musical video regularly featured on Radio Television of Serbia, a CD issued by national broadcasting agency (PGP RTS), as well as performances at the opening which was broadcast live from the scene. The release of another CD was planned, for which there was to be a poetry competition for schools and prominent musical stars were to write the music for selected verses, but the project was never completed. Music videos used in the fund-raising campaign give the first significant clue to how different social groups, defined in terms of musical and political preferences and lifestyle, were encouraged to participate in the project. The music video *Avalski toranj* ('Avala TV Tower') used to promote the rebuilding of Avala TV Tower, produced by RTS, features Bora Đorđević and Dragoljub Đuričić, both open critics of the Milošević's regime in the 1990s. Also present, was a younger star Željko Joksimović who owes most of his fame to the 2004 Eurovision Song Contest and is perceived as an apolitical person. The music video bears an uncanny resemblance to the concerts organised during the bombing, as the location where the video was filmed is the ruins of the former TV tower emphasising the circumstances and aftermath of the NATO campaign (see Figure 3.1). The music composed by Kornelije Kovač, a well-known pop and rock composer, fuses important features of rock, thus appealing to the urban young and middle-aged population (Cvetičanin 2007; Dragićević-Šešić 1994). The lyrics were written by Ljubivoje Ršumović, a famous children's poet, again perceived as an apolitical person. However, as can be concluded through an analysis of its rhetoric, the verses are far from apolitical even though they cling to a banal, ambivalent blend of commonplace tropes:

> The godless devils have managed
> to separate us from the world.
> They destroyed the three-legged tower,
> Belgrade remained without its flag.
> [...]
> We want a life without barriers
> and the life wants us to be free!
> This lightning rod over Belgrade
> Defends us from the world's loneliness

While the first stanza clearly refers to the NATO bombing, and mobilises negative emotions connected to this event, the third stanza portrays the Serbian nation as a modern state integrated within the international community. The trope of a 'life without barriers' is also a clear reference to anti-Milošević rhetoric used in 2000. The song featured clear, stereotypical, one might say vacuous references to rock music, such as the prominent role of the electric guitar and strong drum beats, which was also supported by the visual and performative level in the music video. These features, once a cachet of free expression, were here relocated into the new ideological system of post-Milošević Serbia, where rebuilding projects such as this one were not seen as overtly nationalistic, but as depoliticised or simply 'patriotic' (Figure 4.1).

The CD that accompanied the project and the fund-raising campaign was made to appeal to the widest possible audience, but again in a way

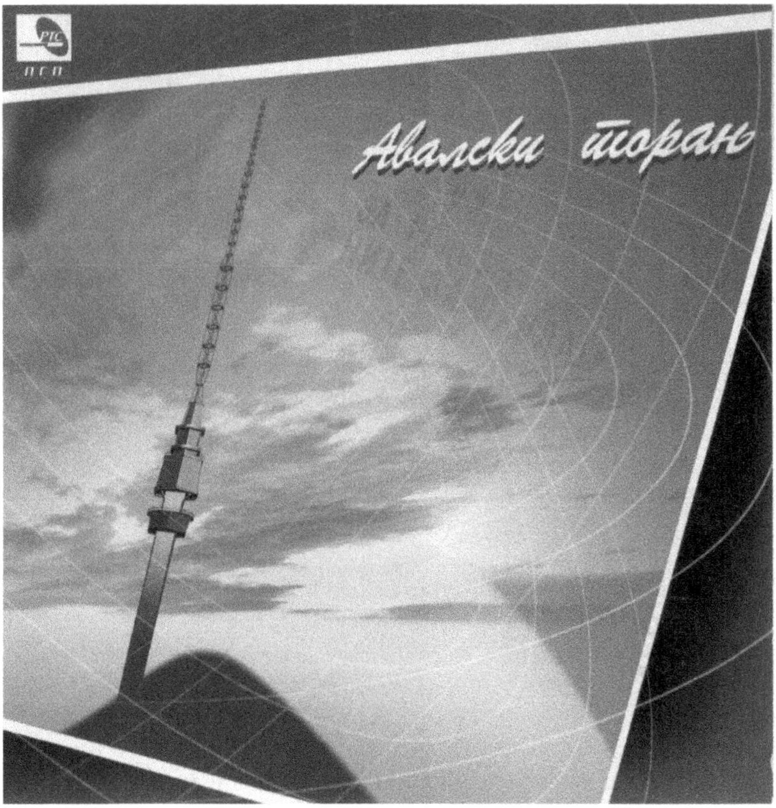

Figure 4.1 Cover of the CD *Avalaski toranj* (Avala tower), 2006

to reflect the odium towards folk music. The compilation was made from popular songs by musicians who agreed to join the fund-raising campaign, some of them appearing as performers on the previously discussed single which also figures as the title number on this CD. Musicians listed on this compilation again present a blend of well-known artists who were either apolitical or had openly allied with the anti-Milošević cause (in this respect, an important addition to Đorđević and Đuričić is the name of Đorđe Balašević). Interestingly, a number of performers connect this CD with the one issued by Target Rekords during the bombing. By analysing the songs, it can be concluded that mainstream rock performers make up most of the artists, but the songs 15 chosen from their repertoire to be included in the compilation were closer to pop music in style, appealing to wider audiences and not only to rock music lovers. In this way, the symbolic capital of rock music, as the 'Western' genre, rebellious, original and individualised, was preserved, while the compilation, at the same time could please the widest possible audience. As the March 1999 concert in Belgrade shows, rock music was relocated and recontextualised as an instrument of banal Serbian nationalism, supporting seemingly innocuous rebuilding projects, which nevertheless had a strong territorial agenda. In this example through the processes of relocation the impact and aesthetic consequences were much stronger. The banalisation of the nation entailed the banalisation of the rock music itself, which was further accentuated with additions of songs performed by pop musicians from the pre-rock era as well as children's pop songs.

Conclusion

In his study, Billig scrutinised the case of Western, first world countries, which do not usually feature in studies of nationalism. However, studying popular music as the instrument of banalisation implies that there is something intrinsically 'Western' in nationalism as the dominant governing ideology of the globalised world. Not only did the modern ideology of the nationalism originally develop in the advanced countries of the West, but also the 'progressive' and desirable forms of nationalism, are continuously perceived as something that should be modelled on successful Western nations, emulating their cultural patterns. In order to impose their authority, the Federal Republic of Yugoslavia and Republic of Serbia recycled almost exclusively Western-styled popular music which was originally perceived as non- or even anti-nationalistic, transmitting the modernising values of the Western world. Western-influenced popular music was relocated from its original context and ascribed with new meanings of banal patriotism. Relocated into the newly formulated state campaigns, it was used in order to mollify the 'hot' nationalistic practices and to attract a wider audience consisting of all generations. It seems like, while all hot nationalisms are hot in their own way, the only way to create banal nationalism is by adopting and recycling products of popular culture in general, and music in particular, by which a nation-state can become widely recognised as Western, modern, and progressive.

The two campaigns described in this chapter bear many resemblances: a CD was produced saturated with recycled Western-styled popular music; an emblematic music video was specifically produced and continually screened on state television; an array of rock and 16 pop rock musicians was mobilised in a seemingly unproblematic cause. The importance of the banality of everyday experience is affirmed in both cases: the two campaigns were both in a way a response to the disruption of the everyday experience of the population, either by the 'sonic war' of NATO in Belgrade, or by the pronounced absence of the TV tower that the citizens of the capital were used of seeing in the landscape. In both cases, campaigns struggled to make changes in the realm of everyday, either by staging a public concert as a counterweight to the sonic war, or by restoring the all too familiar vista. As dictionaries do not define banal simply as mundane, unmarked, but specifically insist on the quality of lacking originality and the complete absence of novelty and freshness, these examples also show that banal nationalism is similarly not something which just happens every day or which has lost all immediate appeal.[6] For practices of banal nationalism to function and truly serve their purpose of reaffirming nation-state authority, they have to be based on cultural practices which themselves lack originality, which are recycled relics of popular practices or products which have either lost or changed their original purpose. The Western-styled popular music used in this state campaign has thus not only lost its originality, but has itself become banalised as a part of the state apparatus, and its original meaning as a type of music which allows 'freedom of expression' has been lost through the described processes of relocation and its submission to the state apparatus. The genres such as rock and roll, which had previously served as markers of individual freedom, have now been relocated into the new political post-socialist context of the Serbian nation-state struggling to affirm its authority. Even more specifically, they were relocated onto the very sights of national traumatic experiences – the urban spaces as the potential targets of NATO bombing and the devastated ruins of sites which had been bombed – and used to banalise these very sites and the sonic and visual trauma connected to them.[7] Finally, by relocating these popular music genres from their original context into new spatiotemporal realities, they not only lost part of their aesthetic appeal, but they also become divested of their political purpose.

Acknowledgements

The chapter was written as a part of the project Serbian Musical Identities within Local and Global Frameworks: Traditions, Changes, Challenges (no. 177004 /2011–2014/) funded by the Ministry of Education and Science of Republic of Serbia.

Notes

1. The importance of Billing's contribution notwithstanding, in recent years scholars have been re-evaluating the impact of his approach on nationalism studies. Craig Calhoun (2014) rightfully concludes that the applications of Billing's banal nationalism were mostly about retracing the steps of Billing's research and providing an array of case studies reaffirming his argument, rather than dealing with the important consequences of his conclusions. Calhoun notices that by merely describing how 'the world of nations' is reaffirmed through banal practices, not only do we fail to come closer to the answer to how it got formed in the first place, but we can even become complicit in its making. The issue was also addressed by Rhys Jones and Peter Merriman, who note that Billig's account also implies a 'separation between the banal and hot processes that reproduce nationalism' (Jones and Merriman 2009: 165), and try to put emphasis on the notion of everyday nationalism, whereas certain practices can be interpreted both as 'banal' and 'hot', depending on the wider social and discursive context.
2. The Dayton Agreement signed in November 1995 (Slobodan Milošević being one of the signatories) brought the end to the conflict in Bosnia and Herzegovina, several months after fighting in Croatia also ended. In the following year, Federative Republic of Yugoslavia (consisting of Serbia and Montenegro) recognised the sovereignty of Croatia and Bosnia-Herzegovina. Using the term 'post-Dayton Serbia' I wish to emphasise that silent disavowal of its role in these conflicts (and in encouragement of the belligerent Serbian nationalism in general) is one of the constants in Serbian state politics from the Dayton Agreement to this day, and that it runs through both the Milošević and post-Milošević periods.
3. Cf. Appeal Judgement Summary for Šainović et al., International Criminal Tribunal for the former Yugoslavia (2014) http://www.icty.org/x/cases/milutinovic/acjug/en/140123_summary.pdf.
4. Quite different from the communist Socialist Federal Republic of Yugoslavia, the Federal Republic of Yugoslavia, founded in 1992, consisted only of Serbia and Montenegro, the republics which did not choose to leave the combined Yugoslav state (unlike Slovenia, Croatia, Bosnia-Herzegovina and Macedonia). In this period, its government was strongly associated with Milošević's regime. He acted as the federation's president between 1997 and2000. In 2003, this polity changed its name to to the State Union of Serbia and Montenegro and finally ceased to exist with secession of Montenegro in 2006.
5. This is not to claim that there were no casualties, and NATO's official naming of these with the term 'collateral damage' was received with indignation. Particularly exposed by media was the death of three-year-old Milica Rakić who was killed by shrapnel during the bombing of a military facility in the outskirts of Belgrade and the most notorious was the death of 16 workers at Radio Television of Serbia when its building was bombed on 23 April. The director of television, Dragoljub Milanović, was later convicted for not evacuating people from the building, and the families of the victims expressed their belief that they served as victims to bring international attention to unjustified targets in the NATO bombing campaign (Vasić 2012). During the bombing (and afterwards) Belgrade residents remained largely oblivious to the simultaneous crimes of government forces in Kosovo.

6. The etymology of banal provides an equally useful tale: stemming from the mid-eighteenth century, it was originally related to feudal service and meant what was 'compulsory' and hence 'common to all' (Hoad 1996). Not surprisingly, banal nationalistic practices serve the purpose of inscribing the compulsory 'national identity' onto the population of the state, reaffirming the authority of the nation-state and enabling the governing of the population.
7. The analysis of the campaigns also provides us with an important insight pertaining to the ways in which Serbian nationalism deals with recent conflicts. Namely, both campaigns were directly aimed at hiding the belligerent nature of Serbian nationalism and the wry picture of the 1999 conflict can be construed as a part of the banalised narrative advocated by the Serbian state. It excludes the insight into the reasons which led to the NATO bombing and information on the actions of Serbian troops towards the Albanian population in Kosovo. Not only was this the case in 1999, but it continues to be so even in the present day, when the everyday landscape of Belgrade with its flagship Avala TV Tower was restored.

Works cited

'Appeal Judgement Summary for Šainović et al., International Criminal Tribunal for the former Yugoslavia' (2014). http://www.icty.org/x/cases/milutinovic/acjug/en/ 140123_summary.pdf (accessed 31 August 2014)

Arnautović, Jelena (2012). *Između politike i tržišta. Popularna muzika na Radio-Beogradu u SFRJ* (Beograd: Radio-televizija Srbije).

Atanasovski, Srđan (2012). 'Turbo-folk as "bad music": Politics of musical valuing', in Sara R. Falke and Katharina Wisotzki (eds), *Böse Macht Musik. Zur Ästhetik des Bösen in der Musik* (Bielefeld: Transcript), pp. 157–172.

Bhabha, Homi K. (1994). *The Location of Culture* (London and New York: Routledge).

Billig, Michael (1995). *Banal Nationalism* (London: Sage).

Calic, Marie-Janine (2010). *Geschichte Jugoslawiens im 20. Jahrhundert* (München: C.H.Beck)

Calhoun, Craig (2014). Keynote address at the *Everyday Nationhood* symposium, 8th September, Birkbeck College, University of London. http://youtube/aS1UV4LQj1Q (accessed 31 August 2014)

Cvetičanin, Predrag (2007). *Kulturne potrebe, navike i ukus građana Srbije i Makedonije* (Niš: Odbor za građansku inicijativu).

Doknić, Branka (2013). *Kulturna politika Jugoslavije 1946–1963* (Beograd: Službeni glasnik).

Đorđević, Radomir (2002). *Leksikon potkultura* (Niš: Zograf)

Dragićević-Šešić, Milena (1994). *Neofolk kultura: Publika i njene zvezde* (Sremski Karlovci and Novi Sad: Izdavačka knjižarnica Zorana Stojanovića).

Đurković, Miša. (2001). 'Ideologizacija turbo-folka', *Kultura* 102, pp. 19–33.

Gledhill, Christine (ed.) (1991). *Stardom: Industry of Desire* (London and New York: Routledge).

Goodman, Steve (2010). *Sonic Warfare: Sound, Affect, and the Ecology of Fear* (Cambridge, MA: MIT Press).

Gordy, Eric D. (1999) *The Culture of Power in Serbia: Nationalism and the Destruction of Alternatives* (University Park, PA: Pennsylvania State University Press).

Grujičić, Nebojša (1999) 'Priče u senci rata (3). Ko to tamo peva?', *Vreme* 443, 3 July 1999. http://www.vreme.com/arhiva_html/443/10.html (accessed 31 August 2014)

Hoad, T. F., ed. (1996). *The Concise Oxford Dictionary of English Etymology* (Oxford: Oxford University Press).

Ivačković, Ivan (2013). *Kako smo propevali. Jugoslavija i njena muzika* (Beograd: Laguna).

Jones, Rhys, and Peter Merriman (2009). 'Hot, banal and everyday nationalism: Bilingual road signs in Wales', *Political Geography* 28, pp. 164–173.

Kos, Koraljka (1972). 'New dimensions in folk music: A contribution to the study of musical tastes in contemporary Yugoslav society', *International Review of the Aesthetics and Sociology of Music* 3, pp. 61–73.

Kronja, Ivana (2000). *Smrtonosni sjaj: masovna psihologija i estetika turbo-folka* (Beograd: Tehnokratija).

Kupres, Radovan (2004). *Sav taj folk*, TV documentary (Belgrade: Television B92).

Longinović, Tomislav (2001). 'Music wars: Blood and song at the end of Yugoslavia', in Ronald M. Radano and Philip V. Bohlman (eds), *Music and the Racial Imagination* (Chicago and London: University of Chicago Press), pp. 622–643.

Marković, Predrag J. (1996). *Beograd između istoka i zapada 1948–1965* (Beograd: Službeni list SRJ).

Mijatović, Brana (2008). ' "Throwing Stones at the System": Rock Music in Serbia during the 1990s,' *Music & Politics* 2. http://dx.doi.org/10.3998/mp.9460447.0002.203 (accessed 31 August 2014)

Rasmussen, Ljerka V. (2012). *Newly Composed Folk Music of Yugoslavia* (New York and London: Routledge).

Sundhaussen, Holm (2007). *Geschichte Serbiens. 19–21. Jahrhundert* (Wien, Köln and Weimar: Böhlau Verlag).

Vasić, Biljana (2012). 'Sto godina samoće', *Vreme* (1131), 6 September 2012. http://www.vreme.com/cms/view.php?id=1070972 (accessed 31 August 2014).

Vučetić, Radina (2006). 'Rokenrol na Zapadu Istoka. Slučaj Džuboks', *Godišnjak za društvenu istoriju* 13, pp. 71–88.

Vučetić, Radina (2012). *Koka-kola socijalizam* (Beograd: Službeni glasnik).

Yack, Bernard (2012). *Nationalism and the Moral Psychology of Community* (Chicago and London: University of Chicago Press).

5

'Escape and Build another World': Relocations in Classical Minimalism and Minimal Techno

Isabel Stoppani de Berrié

In this chapter I discuss the relationship between two musical minimalisms: the minimal techno which developed in Detroit during the early 1990s and the more 'classical' form of minimalism belonging to composers in downtown New York, especially Manhattan, during the 1960s. Initially, by returning to the early manifestations of each, now seen as 'classics,' I demonstrate how these two traditions intertwine at their roots, before examining their subsequent interaction, as more mature and diverse artforms, in the 1990s and later. By looking at the reworkings by techno DJs of classical minimalist works, and vice versa, I aim to bring into relief cultural differences between the different performance media, spaces, and geographical heritages of each. I document the tension between the universalist hopes of minimalist artists of various kinds, on the one hand, and the persistence of genre distinctions on the other. The latter remain relevant due to the continuing use of critical concepts which contribute to the maintenance of hierarchical boundaries between the classical and techno spheres and corresponding modes of consumption.

A brief history of musical minimalism from local roots to global phenomenon

At the time of writing, both minimal techno and classical minimalism have superseded their original distinctive impulses, giving way to globalisation and spawning a range of subgenres. During the 1980s and 1990s, a new generation of minimalists were embracing a new kind of minimalism which the American composer and scholar Kyle Gann has called 'totalism' (Gann 2006) – a kind of music focused on rhythmic complexity, with polyrhythmic tendencies and a motoric aspect.

Minimal techno, which began in Detroit as a move back towards the repetitive basics of the earliest Detroit techno of the 1980s, enjoyed a boom in Europe, particularly Germany, during the 2000s. Both of those subsequent scenes have by now died down; mainstream attention has deserted minimal techno, (Naylor 2008) and the composers associated with totalism have shifted their focus to other techniques. Today, a late stage of diversification sees minimalist influence reflected across genres. Meanwhile, the pioneers of classical minimalism – Steve Reich, Philip Glass, Terry Riley – are enjoying greater audiences than ever before, and there is considerable recognition of the 'Belleville three' credited as the originators of Detroit techno: Juan Atkins, Derrick May, and Kevin Saunderson. Many of these figures continue to create new music in addition to giving many interviews in which they reflect on their own past work.

Part of this change has been a progression from the radical simplicity of early examples towards greater complexity, along with a loss of the stylistic purity attributable to the smaller number of original practitioners. The time has passed for a definition of musical minimalism in which, in Edward Strickland's words, 'overt and immediately audible repetition of simple, even simplistic material is the predominant structural principle' (Strickland 1993: 13). Kyle Gann has made the bold claim that minimalism is the acorn from which a whole new musical language, rather than just a specific genre or set of subgenres, can grow. Meanwhile, the Detroit-raised techno DJ Magda commented, on being asked about the mainstream nature of minimal techno in 2008, 'minimal techno doesn't have to be 'ticka-ticka-ticka-ticka' (Raven 2008). This sense of freedom, and of the looseness of genre boundaries, plus, sometimes, the fringe nature of the latest outgrowths, contributes to the sense of possibility regarding cross-pollination. Also (as I shall later explain in greater detail), the political idea of minimalism as a universalist, global phenomenon is part of its utopian ideal and forms part of the marketing approach of record labels and promoters as much as it is a sincerely-held belief among composers and DJs.

To place the international mélanges of minimalism in context, I will first construct a picture of their intensely local beginnings. Each of these movements was famously located within a particular culture in which geography is considered a central element of the definition of the culture, both at the time, and in later histories. In both print interviews and in a number of documentaries, some of the key figures of early Detroit techno have commented retrospectively on their working environments and discussed how the environment shaped the music.

This has now become part of the origin myth and mystery of Detroit techno, and something which gives it romance in the popular imagination. The imagery and commentary of documentary-makers often serve to reinforce this narrative (Figure 5.1).

Detroit Techno City, for instance, opens with images of the run-down city and industrial decay. In the documentary *Universal Techno*, a thesis is put forward according to which an early, localised scene gave way to an international musical language. Sights such as a destroyed cinema, now a car park, are also displayed (Universal Techno, 14.22). Factors cited as having influenced the musical developments include that there wasn't much to do in Detroit (23.00), that it was affected by the technological revolution (which led to the decline of work available in the Ford car factory, which had been a central part of city life) (19.00) and that industry is the focus of everyone in Detroit, so that people end up having a very unaffectionate, cold affect (28.00). Of these, the effects of the technological revolution, and subsequent industrial decay following automation, are foremost. Carl Craig, a significant figure in later Detroit techno, claims 'we've seen the destruction of the city, and we've been able to relate that to what we've done with our music' (4.57). The producer Robert Hood, who was one of the early contributors to the development of a minimal techno style, describes how specific aspects of the experience of living in Detroit, in a sensory way, could have been translated synaesthetically into music,

> Back in '92, '93, I sort of crafted this sound I liked to call 'the grey area.' What that is is the atmosphere in Detroit from the factories

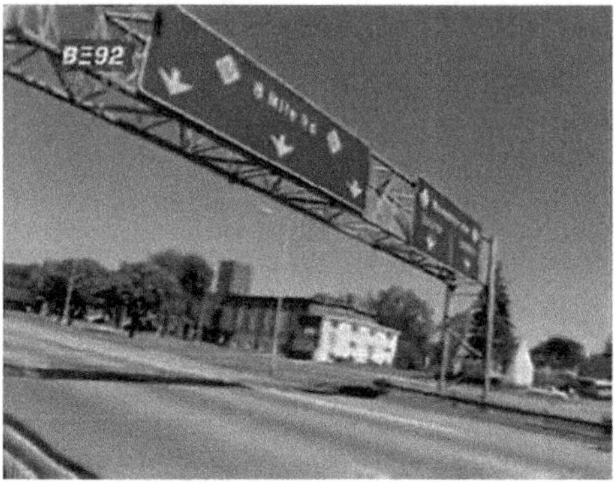

Figure 5.1 Eight-mile road, which marks the boundary between Detroit and the suburbs, still from *Detroit Techno City*

> that emit this smoky and grey climate, this atmosphere. It's just like ash and dirt from the sewers that just come up from the car emissions and it creates a grey haze over the city. That inspired me, and I kind of piggybacked off of what my contemporaries had created and rode along with that and interpreted my own Detroit sound from that. (Stolman 2012)

In academic literature, too, the influence of specific aspects of the Detroit environment on the formation of techno has been well-referenced. Carla Vecchiola (Vecchiola 2011) writes about the Submerged techno label (which was started towards the end of that early period) characterising techno as a creative response to urban crisis. Deborah Che (2009) seeks to demonstrate that techno could not have been born in any other city but post-Fordist Detroit. She references the high disposable income provided by Fordist production which gave automotive factory workers the money to spend on Motown records. This Motown sound was an important component of techno, in addition to the European influence of groups such as Kraftwerk and Tangerine Dream, German electronic groups whose machine-like sounds resonated with the Detroit engagement with technology. Hence, techno has been described as 'machine soul' (Savage 1993) and 'hi-tech soul' (as in the documentary of that name). Perspectives on the origins of techno have also been advanced by Simon Reynolds (Reynolds 2008) – a more autobiographical account – and Dan Sicko (Sicko 2010) also demonstrates broad agreement with this view.

By comparison, suggestions regarding the urban environment's role in the development of classical minimalism are less detailed and more speculative. Tim Lawrence, in his article on Alfred Russell, a genre-crossing composer who shared the city with classical minimalists in downtown New York (1973–1992), uses the rhizome idea in an attempt to draw attention to 'a body of interacting buildings, creative producers, technologies and other components' (Lawrence 2009: 6) rather than centring the music or the composer. His approach gives space and buildings an active role. Lawrence points out that the area of Manhattan he discusses was previously home to the manufacturing of lights – when this ended, the empty industrial buildings provided space for artists. He argues that 'the expansive contours of the lofts inspired them to develop big, bold, energized works – works they might not have produced in another milieu' (Lawrence 2009: 9). This picture of a previously buzzing industry giving over its empty spaces to music invites a parallel with the situation in Detroit. On a more detailed level, classical minimalism's musical verticality (vertical time) has been sourced directly to the vertical architecture of Manhattan by Kiran Sande (Sande 2010). Thus, the idea of exchange and communication between different kinds of music, the crossing of boundaries (between music and art, between classical and popular, between uptown and downtown) is intrinsically interlinked with the ways in which the music is perceived as stemming from and communicating with its place of origin.

Many of Steve Reich's pieces involve local sounds which reflect the atmosphere of the places in which they are recorded – for instance, the San Francisco street preacher of *It's Gonna Rain* and the young man of *Come Out*, who has been beaten but won't get taken to hospital unless he makes his bruises bleed. Reich's recent work *City Life* incorporates many of the sounds of New York: car horns, doors slamming, air brakes, subway chimes, pile driver, car alarms, heartbeats, boat horns, buoys, fire and police sirens.[1] Reich, in 2006, gave his own perspective on the use of place in music,

> ... *Vermont Counterpoint* was written largely in New York, *New York Counterpoint* was written largely in Vermont, and *Tehillim* was written partially in New York, partially in Vermont, and partially in Stuttgart. I've learned over the years that geography is not that important.... But all music does come from a time and place. I was born and raised in New York. We have recently just moved out of New York, but it's inside of me and it will be inside of me until they put me in a box in the ground. [...] Really great stuff is local. It's very rooted in a particular time and a particular place. It's just that local factor that, strangely enough, makes it universal. At least, that's what I believe. (Klein 2006)

Reich's music is rooted not only in the particular time and place to which he refers, but also in his personal and subjective curation and transformation of the sounds within his orbit. This construction, in turn, has the power to influence subsequent responses to the sounds of the city, and influence the impressions of New York for locals and visitors alike.

With respect to both minimal techno and classical minimalism, the urban environment does not influence the music in a one-sided way. In addition to music being born of a place, music also inhabits and colours that space and in some ways also creates that place, or creates virtual space. Urban space calls out for music to colour and shape it – the more so, perhaps, if it seems to act as a blank canvas or to be so grim as to propel its inhabitants to seek transformation or transcendence. This idea is referenced in *Detroit Techno City* – citing the American composer Morton Feldman's *Rothko Chapel* as an example (thus already drawing a point of connection between DJs and art composers). Furthermore, through music, space can be transformed yet further, as Robert Hood, one of the first minimal techno producers, describes, the desire to escape 'it has been, for me as a techno artist, a vehicle, a spaceship to escape – at least creatively, artistically – and build another world, another reality.' This tallies with the futurism of techno, and a drive towards the future as an attempt to escape the run-down present. Philip Glass has also noted the power of minimalist music to construct place, rather than merely reflect it, remarking in an interview, 'Music is a place. It's a place which is as real as Chicago, or Sydney, or any place you've ever been to. It's an absolute actual place' (interview with Brian Ritchie, 6.00).

Two models, transplanted elements

Having established strong local components to minimalist compositions, I shall now identify two non-mutually-exclusive models which underlie accounts in academic and journalistic literature of the relationship between classical and non-classical kinds of minimalist music. Both models are suggestive of interpretations of minimalist music-making as a process which involves the integration of elements transplanted from another culture, whether from Europe, America, or further afield. Therefore, when I discuss the more recent dialogues between composers and DJs, it is against a background of ambiguity in which the to-and-fro of minimalist cultures, and the analytical dissection of all music into minimalist components by practitioners on both sides of the divide, leaves open the possibility that even in their most provocative encounters, minimalist musical cultures are only re-encountering variant versions of themselves, filtered through nominal hierarchical oppositions of classical/popular, American/European, and black/white.

The first scenario, which I shall term the 'vertical model,' focuses on the formative influence of classical minimalism on the origins of minimal techno. In terms which may appear pejorative, Daniel Chamberlin suggests that minimal techno lifted substantially from classical minimalism; citing early minimal techno artists Basic Channel, Richie Hawtin, Wolfgang Voigt and Surgeon and claiming that, 'All of them take the phase-shifting percussion from Reich's *Clapping Music*; crib sine wave drones from La Monte Young's[2] *Drift Study*; and above all, repeat the single-minded patterns of Riley's *In C'* (Chamberlin 2003). Philip Sherburne has identified three ways in which techno producers make use of minimalist music: sampling, emulating, and remixing (Cox and Warner 2004: xiv).[3] Christoph Cox claims that, 'the kind of layered, modular repetition fostered by Reich and Glass is the stuff of which techno is made' (Cox 1999).[4] Reich's early tape pieces, *Come Out* and *It's Gonna Rain*, are also very frequently cited as influences. Terry Riley's all-night solo harmonium performances of the early 1960s have also been identified as 'an obvious precursor of the all-night underground raves to follow decades later' (Ankeny). La Monte Young's *The Well-Tuned Piano* (starting from 1964), which was also accompanied by coloured lighting, provides another such example.

Some audible similarities could be interpreted as evidence of Reich's direct influence – take, for example, *Spastik* (a 1993 track by Ritchie Hawkin under the name of Plastikman) and compare with Steve Reich's *Drumming*. For a recent example of something resembling Reich's early tape pieces, we might consider Thomas Brinkmann's *Ikaria* (2004). The music critic Mark Richardson, in a Pitchfork review, commented on the resemblance of this to Reich's early work. Brinkmann also uses looping voices in other tracks, such as *Ribosom* (2002) – in which Brinkman reads out scientific information over a minimal backing. However, I would argue that *Ikaria* is remarkable and worthy of comment precisely because it is rare to hear in mimimalist techno anything so overtly similar to Reich's early tape pieces. There is little musical evidence of such specific influence.

While some continuities may appear suggestive, Tim Rutherford-Johnson (2013) has recently commented on the problematic nature of the ways in which narratives concerning Reich's influence, in particular, have been perpetuated. He suggests that part of the reason for the survival of this narrative is that it is beneficial to those involved. For the classical minimalist composers still alive, it bolsters their claims to possess a significant legacy. The potential for minimalism to become a dominant musical language is highlighted, and it is credited with healing the rift between classical music and its audience (helping minimalism to win the battle against other forms of modernist music such as serialism, seen as elitist and incomprehensible). The critic Alex Ross has claimed, 'We are living in a world scored by Reich' (Ross 2006). Similarly weighty claims have been made for Terry Riley's *In C*, originally chosen by the record company for its crossover appeal. According to David Lang, '*In C* was the same kind of life changing work that *The Rite of Spring* had been to an earlier generation of composers' (Oteri 2010). For techno artists, meanwhile – as Rutherford-Johnson argues – the prestige of the classical music world can help them to acquire some intellectual and high-art credentials. This reflects a hierarchical view in which popular music is seen as debased recycling of high-art elements.

According to the second scenario, which I term the 'horizontal model,' it is implied that different kinds of musical minimalism, rather than being influenced directly by one another, share common influences which can account for perceived similarities. The music journalist Philip Sherburne (2004) emphasises that both minimal techno and classical minimalism are inspired by African rhythm and they focus on rhythm rather than melody or progression. Both, also, were in dialogue with Europe – classical minimalism in a conscious desire to abandon European modernism, early Detroit techno through the import of synthesiser sounds and a German industrial aesthetic – which gave the music, initially, a sound which was new in Detroit.[5] In *Detroit the Blueprint of Techno,* techno's non-native aspect is emphasised. Derrick May says of the Europeans who later appropriated and reified the Detroit sound – leading to complaints of exploitation because of the lack of recompense for the originators who were mimicked – 'maybe, subconsciously, we made music more for them than for us' (Detroit the Blueprint of Techno, 8.11). He also says he's never really had the support of the community (10.15). An early listener recounts how he was surprised to find out the music was local, and not European as he at first assumed (11.32). Thus, both classical minimalism and minimal were initially defined in opposition to European predecessors, before later being widely exported and re-appropriated in order to satisfy the European appetite for novelty.

Others have suggested that a wide range of musicians have been independently affected by new technological developments and the possibilities afforded for the manipulation of sound, e.g. the ability to produce exact repetitions. Cross-pollination with visual art was another key feature of the development of classical minimalism. A strong connection with visual art is less often mentioned in relation to minimal techno, but since 'minimalism' was originally a term borrowed from the art world,[6] it is possible that a range of musicians shared artistic inspiration as a source. Thomas Brinkmann's techno record label is named after the visual artist Max Ernst, perhaps because his 'frottage' or rubbing technique is linked with the way in which DJs use vinyl.

Taking into account these many common sources, it's possible to see the techniques and materials of minimalism as belonging to a general pool, a tool-kit crossing genre dividing from which anyone can take anything – and Sherburne takes this view, regarding tracks as simply repositories for raw material for DIY DJ culture. Robert Fink emphasises the technological aspect by stating that,

> Ambient needs a music scholar willing to research as if there were no canonic boundaries, to see that for electronic musicians there is only one distinction that matters. There are people who use tape and electronics ... and people who don't (2002).

It may be that Fink imagines technology as already implying method and style through its intrinsic possibilities. Repetition – at its most extreme in early classical minimalism and in minimalist Detroit techno (as in the very earliest Detroit techno) – is one of the most basic processes through which material of diverse origins can be subjected to equal treatment. Once within this sphere, and subject to the democratizing force of technology, the various elements may be seen as equalized; this appears to open up the corresponding possibility of human beings existing together on the same level. But despite its allure, the vision of an ambient economy of freely exchanged sounds rests on the divorce of sounds from their origins – a divorce which I suggest is rarely complete or simple.[7] It also suggests a meeting of classical minimalism and minimal techno on more equal terms. In the next section I will challenge aspects of both the 'horizontal' and 'vertical' models, neither of which does justice to the ways in which sampled material, transplanted and processed, continues to evoke for audiences its original contexts.

Interactions

Following the increased global spread of minimalism, musical minimalism, like the related category of 'ambient' music, has sometimes been positioned as defying divisions between classical and non-classical music, representing a new era in which old constructions of genre have lost their relevance. Ben Neill, composer of post-minimalist music, commented in 2000 that the boundary between classical and non-classical music had disintegrated. Of the release of *Reich Remixed* (1999), a collection of DJ remixes of music by Steve Reich, the notes provided by Nonesuch Records (listed on their website) boasted,

> It is the prevailing relevance of Reich's music to the post-modern chill-out rooms and concert halls alike that now lead us to *Reich Remixed*.... In addition to paying homage to Reich's work, *Reich Remixed* breaks

down previously exclusive boundaries between classical and club music to create a new sound for audiences of today and tomorrow.

The processes of translation required for the migration of material across boundaries serves to confirm the continuing existence of those boundaries and their power of definition over the materials they encompass. Thus, I aim to demonstrate that the classical/popular distinction, along with other genre divisions, remains relevant in minimalist music, despite suggestions to the contrary by scholars, record companies, and composers. I focus on certain aspects which may require translation between different minimalist cultures: the intersecting questions of medium (live vs. electronic) and technique, in particular how repetition is used and the extent to which form is understood as arising from the sound material, rather than being an independently-determined process. The lens through which I bring these themes into focus will be that of the reworking of classical works by techno artists and vice versa.

Virtuosity and the sources of sound

In the past, techno has tended to be markedly technological in the origin and organisation of its sounds, while classical minimalism has involved predominantly live instruments. These latter have often been those traditionally used in the Western concert hall, though often in idiosyncratic combinations and on a scale smaller than the orchestral. There are of course early exceptions: the use of some sampled sound in classical minimalism (particularly by Reich) and the appearance of samples from the Detroit Symphony Orchestra on the Derrick May track *Strings of Life*. Instrumentation is one of the most audible differences between classical and techno: Christoph Cox contrasts the 'marimbas and bongos' of classical minimalism with the 'samplers and sequencers' of techno but only in order to point out that the music is structurally similar (Cox 1999). Despite the optimistic aspects of technology, the Detroit DJs' situation also involved a kind of deprivation – they did not have access to the sleek studios with which German producers would later produce Detroit-inspired tracks. Instead, they made do with relatively simple equipment. Classical minimalists, also, were working with limited resources in terms of available instrumentalists, and since their work did not at that time have large uptake by other groups, performances were mainly by dedicated ensembles. This placed them outside the classical mainstream.

Within classical minimalism, the sound is not always as closely determined at the level of the soundwave; the specificity of a timbre is less crucial in defining a work's identity. Iconic sounds in minimal techno, such as the hoover sound of Human Resource's *The Dominator*, have greater unique identity. Sampling often involves placing the exact soundwave (with or without transformation) into another work. By contrast, the potential uniqueness of sound in each realisation still partially determines the classical. For instance, Reich's *Electric Counterpoint* can be played on any instrument you like, as can Riley's *In C*. This offers the possibility of wide performance by musicians from a range of backgrounds. Phasing Reich's music on two turntables, Thomas Brinkmann comes close to treating the turntable as an instrument.[8] The removal of traditionally pitched content or instrumental sources – as well as the use of limited sonic material – is something which draws together Brinkmann's latest work with Reich's early tape pieces. Minimal techno resisted the creeping in of pleasant-sounding instrumental samples in favour of a return to uncompromising synthetic sounds which often refused to adhere to a tonal framework.

I will now discuss a variety of instances in which minimalist material has been 'translated' from one medium into another, which has also involved recontextualising music from one place in another kind of performance venue. These examples mostly entail DJs remixing recordings, and composers recreating electronic music with live ensembles. There are also several relatively recent instances of DJs working with live orchestral ensembles. First, an example of techno self-consciously becoming live: the Berlin-based group Brandt Bauer Frick's first album, *Make me Real* (2010) began the play on the production of a sound which was expected to be electronic but turned out to be live. The description of their performance on the Southbank Centre website included such terms as 'genre-defying' and 'successfully fus[ing] techno and classical music.' 'Like minimalist classical ensembles (Steve Reich, Graham Fitkin), they hammer away at tuned percussion and keyboards (including Cagean prepared piano), but there's a difference in inflection: they rock' (Walters 2011). Steve Reich has also hailed this development in the ability of classically-trained musicians to play music in simple time with a sense of groove (Tingen 1995). Instrumental performance offers virtuosity when techniques do not appear intuitive to human performance. The thrill of difficulty provided by *Piano Phase* being performed by two pianists (creating a challenge) – or even, recently, by one pianist – is echoed by the virtuosic display of Brandt Bauer Frick, playing complex rhythms apparently as accurately as machines (Figure 5.2). Although

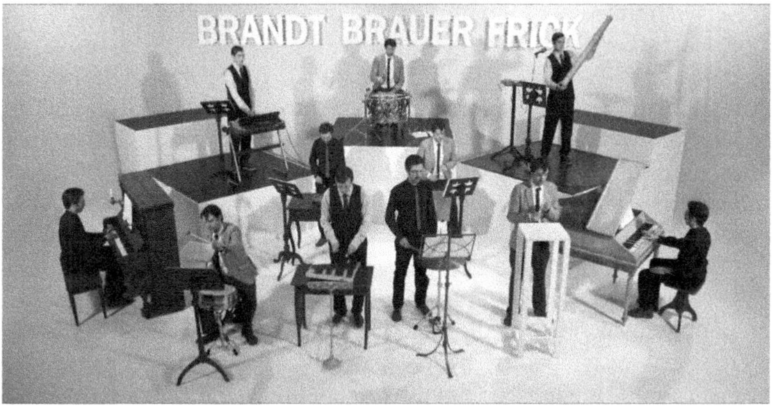

Figure 5.2 Brandt Bauer Frick performing *Bop*

their movements respond to the groove, there is still a robotic aspect to their demeanour which recalls the performance style of Kraftwerk.[9]

The behaviour of the audience also changes depending on what the genre is understood to be. In traditional classical performances, the ensembles move a lot and the audience stays still, while in techno the DJs make more economical movements while the crowd dances. However, sometimes in techno the movement of the DJ has become a part of the spectacle. In the case of the Montpellier Orchestra performance with Jeff Mills, a complaint was that the video doesn't show the charismatic movement of the performer, the virtuosity of his performance as a DJ (Eeles 2006). On the DVD, most of the orchestral players are stiff, but the tuba player can be seen beginning to groove (Blue Potential, 14.48). DJs of the younger generation such as Sven Väth often dance vigorously as they perform. In the case of Brandt Bauer Frick, it has been noted in publicity for their performances that the audience is invited to move – highlighting their dissimilarity to the stuffiness of the traditional classical concert. The audience was also invited to move in the case of the Mills/Montpellier performance, and in a performance of Derrick May's *Strings of Life* with the Macedonian Philharmonic Orchestra they are positively going wild. These performances took place in relatively large venues, with the audience standing.

Because of the necessity for providing parts with musical notation, instrumental arrangements are often carried out not by DJs but by third parties. The iconic electronic musician Aphex Twin has had several reinterpretations for instrumental forces including a version of *Polygon*

Window arranged by Kenneth Hesketh and performed by the London Sinfonietta (a classical chamber orchestra dedicated to the performance of contemporary music) and a version of *Icct Hedral* orchestrated by Philip Glass, the latter taking its cues from the original in terms of sounds already approximating string and wind timbres. It is likely that most classical composers also lack facility with turntables; therefore translations must be carried out, partially at least, by an additional party. Differences of medium are therefore crucial in regulating access to certain performance spaces.

While there is inequality in what the electronic and acoustic modes of performance have to offer, and the expense of a full symphony orchestra is accompanied by prestige, the move from electronic to acoustic is not unequivocally an upgrade, nor is the electronic merely a second-best for when live instruments are not available. Orchestral recreations of techno classics, while gaining warmth, seem to some listeners to lack something in terms of intensity, perhaps the relentlessness of the electronic. Mills, speaking on the *Blue Potential* DVD, seems to equate the use of live instruments with emotion, 'it's possible to put more emotion into electronic music.' Although he embraces this possibility, it does not negate the importance of the 'cold affect' mentioned earlier; the appearance of human warmth in the context of mechanical music gives a particular effect which is different from that of traditional orchestral music – the obverse, perhaps, of the unexpected roboticism of Kraftwerk. Finding these elements out of place, in either case, produces a special *frisson*, through the juxtaposition of the electronic and the organic. Also, the Aphex Twin track involves an example of something which is difficult for a live ensemble to achieve using recognised techniques: the specificity of certain sounds (a chair scraping across the floor, scissors) the usage of which is aligned with the tradition of *musique concrète*.[10] Brandt Bauer Frick have also commented that they needed to modify their instrumental technique by shortening their sounds, and also explored extended techniques in order to be able to approximate the sound palette of techno (Large). In this sense, Reich's use of samplers alongside live instruments is a way to incorporate the possibilities of both sources.

Attitudes to authorship and reworking

Steve Reich's *Radio Rewrite* is a reworking of two Radiohead songs, *Everything in Its Right Place* and *Jigsaw Falling into Place*.[11] Although lending his music to others to remix, before *Radio Rewrite*, Reich had been reticent about incorporating the material of others in his own work.[12] In the interview preceding the radio broadcast of the work, he made it clear that he hadn't used too much direct quotation of Radiohead's music, seeming to be keen on emphasising the status of his piece as original: *Radio Rewrite,* not *Radio Remix.* Thus the classical composer dominates his material. Reich commented 'sometimes you'll hear Radiohead, sometimes you won't.'

One of the criticisms of *Reich Remixed,* a collection of remixes by others of Reich's works, had been that the remixes were too simplistic, simply adding beats but unable to substantially re-work the material – in fact, the album received many negative reviews. Radiohead also had a remix album made of songs from *King of Limbs.* Perhaps anxious to avoid the kind of criticism that had been levelled at *Reich Remixed,* Thom Yorke commented 'I didn't just want floor fillers and all that shit, I just wanted to see how the songs could really branch out and mutate.' Composer Robert Carl, author of a volume on Terry Riley's *In C,* took a similar attitude to remixes and reworkings, 'I like the works that take off from it and actually make truly new pieces ...' (Oteri 2010). Arguably, DJs struggle to subsume classical minimalism into their own idiom because of the new set of expectations adhering to the material. The release of *Reich Remixed* in 2011 placed Radiohead and Reich on an equal footing. This, I suggest, implies that a divide still exists between different spheres of music, an imbalance in the appropriation of material, and the divide is greater between Reich and techno than between Reich and Radiohead – techno producers remix, but Radiohead or Reich must rewrite.

As part of an effort to distinguish reworking from simple appropriation, composers and defenders of classical minimalist reworking have emphasised the learnedness of their practice and the ways in which it casts light on originals. The musicologist Carlo Cenciarelli takes this approach when discussing the ironic and postmodernist aspects of Michael Nyman's reworkings of Mozart. *Mozart 252,* released in 2008, is a whole album of Mozart reworking by Nyman. It includes earlier pieces based on Mozart such as *In Re Don Giovanni* (1977) which uses the first 16 bars of the Catalogue Song from Mozart's opera Don Giovanni, and the soundtrack to *Drowning by Numbers* (1988) which uses material exclusively from Mozart's *Sinfonia Concertante.* Nyman, like the techno artists, has also been criticised for his appropriation of Mozart. The simple method of *Trysting Fields* – identifying each appoggiatura figure and giving it a threefold repetition – can be understood in symbolic terms, since the story of the film is of three parallel tales. Carlo

Cenciarelli acknowledges that *In Re Don Giovanni*, in 'its literal reliance on Mozart's text ... blurs the edges between reception and recomposition.' By bringing to the fore an 'educationalist' aspect, he emphasises Nyman's understanding of the complexities of Mozart's musical language as a prerequisite for his imposition of a later aesthetic. This same knowledge or understanding of the classical original is emphasised in Ivan Hewett's reading of Max Richter's work, in which Hewett compares Richter favourably to non-classical remixers (Hewett 2012). There is also the same concept of presenting Vivaldi (as Cenciarelli claims of Nyman presenting Mozart as a minimalist *avant la lettre*.) Even if – as Cenciarelli thinks – this is ironic in Nyman's case, it is probably not ironic in Richter's.[13]

Yet it is not only the 'classical' minimalists who have engaged in such a way with the musical past. As examples from the (German) techno sphere, one might cite Carl Craig and Moritz Oswald's *ReComposed* which makes use of Ravel's *Boléro*. In this case, it's clearly (as in the title of the work) presented as 'recomposed.' In a sense, this is a double reworking, since Ravel's version of Mussorgsky's *Pictures at an Exhibition* is used, itself a re-orchestration. The record subsequently underwent further remixes. The Villalobos remix *Uli, Mein Ponyhof* is based on the *Gnomus* movement. A short phrase of the movement is excerpted and treated in such a way as to bring out its internal repetition. Therefore, the idea that classical composers have sole claim to an analytic understanding of past music which enables them to remake it intelligently in their own moulds by reclaiming it as 'proto-minimalist' can be shown to rest on shaky ground.

Meanwhile, on the album *Re: ECM*, which he created in collaboration with Max Loderbauer and released in 2012, Villalobos remixed music from the back catalogue of the German record label ECM, which included jazz as well as some Arvo Pärt. In titles, they use the prefix 're-' to characterise each of the tracks. This open-ended usage leaves it to us to categorise it as a remix, a recomposition, or something else. The track *Rekondakion* is based on classical minimalist Arvo Pärt's *Kanon Pokajanen*.[14] The seed of the project was sown when Riccardo Villalobos began playing tracks by Pärt in clubs, beginning with Pärt's *Tabula Rasa* and going on to include music by others. Villalobos's take on the clash of languages, quoted on the ECM record label website, is 'The most important thing is to harmonize these two worlds, without them aspiring to mutually deactivate each other, to keep both – the organic and the electronic – in balance.' Confirming my earlier point concerning the juxtaposition of expressive and mechanical performance elements, this suggests an ecological approach to musical composition. Correspondingly, back in depopulated Detroit, as images of nature reclaiming overgrown buildings become more prevalent, Dan Pitera has developed the concept of a 'syncopated urban landscape' – a term inspired by music – to refer to the possibilities for re-using abandoned buildings and open parkland in a new combination of productivity and greenery (Pitera 2010).

Villalobos and Loderbauer also stated that one of the aims of the project was to make use of the sense of space created by live acoustic recordings. Pärt's *Kondakion* was commissioned specially for the space of Cologne Cathedral, to commemorate the seven hundred and fiftieth anniversary of its building. This indicates another way in which sound samples retain traces of their original location in a way which is relevant to the final expression. By using that acoustic, Villalobos and Loderbauer also access the connotations it brings – the echoes of the European tradition which used that acoustic and the sense of worship. Justin Toland referred to the track as constructing a 'cathedral of sound' (Toland 2011). Videos of orchestral performances of techno also emphasise the setting with images such as the bridges of the night-time city; minimalism remains suited as a soundtrack to cityscapes, adding atmosphere and conjuring the rhythms of urban life. For instance, the opening to *Blue Potential* shows images of both more natural areas and man-made monuments, before moving into alternation between clips of the performance and time-lapse footage of people moving underneath the monumental and historical architectural structures of the city (Figure 5.3).

Processes of recomposition have come to occupy related roles in the classical and popular spheres. It is crucial to the continuation and influence of classical minimalism that one of the features that distinguished its early development – the performance by dedicated groups – was only a stage. The eventual take-up by recognised or prestigious ensembles was crucial to the development of greater awareness. Like repeat performances in the classical sphere, which eventually took the music of early minimalists outside of their dedicated ensembles and into the wider classical music world, remixing – specifically the remix album dedicated to a single track – helps to determine membership of a canon of influential works. In techno, it's been applied to classic tracks such as Terrence Dixon's *Minimalism* (1995) and DBX's *Losing Control* (1994). Perhaps one of the earliest examples, Human Resource's *Dominator* (1991) was released at the time as a 'more than complete' CD of remixes and has been re-released ever since.

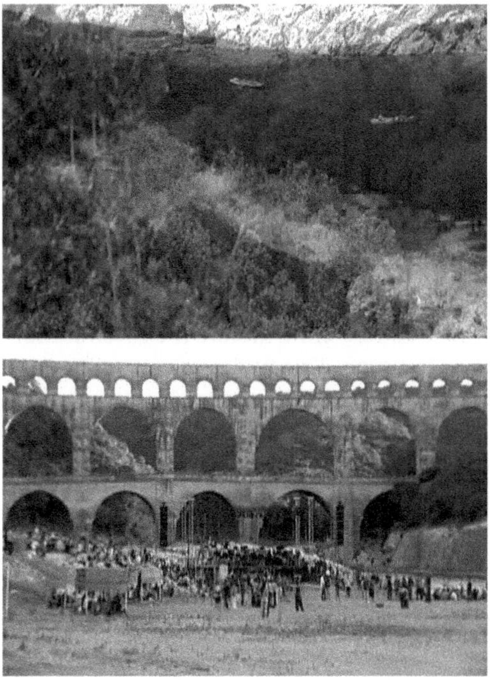

Figure 5.3 People moving underneath monumental and historical architectural structures of the city

In terms of remixes of 'classical' tracks with minimalist qualities, it's often the best-known pieces that receive the remix: Arvo Pärt's *Fratres*, Terry Riley's *In C*, and Michael Nyman's music for *The Piano*. Especially if we take into account bedroom remixes as well as those by household names, the prevalence of remixes of certain minimalist works simply reflects the currency they have gained among a wide audience. However, the remix tribute can also constitute a nod back to origins. Thus, it is no coincidence that two of the pieces mentioned by Daniel Chamberlin as sources have also received dedicated remix albums. Philip Glass would seem to be buying into the DIY remix culture with the release of an app which allows people to remix his tracks themselves. Encouragement from classical composers and record companies could be interpreted in a cynical light, since remixing serves as a way of confirming the continued relevance and influence of a track.[15] The Craig and Oswald work for Deutsche Grammophon, along with the Villalobos/Loderbauer work for ECM, suggests that the German labels saw the potential for using

DJs' skills in order to update their output for a modern audience by imposing a modern aesthetic. The qualities of temporal distance and of geographical distance in source materials are in many ways parallel in terms of their compositional treatment.

Conclusion

Composers and DJs can be understood as fulfilling the role of relocators who, when bringing material from the other sphere within their orbit, transform and reinterpret it in order to bring it into line with generic expectations, while also retaining something of its otherness. The movement of sonic material between classical and techno contexts involves changes in media, performance contexts, and the expectations according to which the work is received. The geographical shift of techno from Detroit warehouses to European venues (both large urban spectacles and intimate quasi-classical concert spaces) brings with it new audiences with different expectations and kinds of participation, access to new realisations (such as the incorporation of orchestral elements) and the varied individual acoustics of resonating spaces. Meanwhile, the sense of place in minimal techno intersects with that of time; conjuring of the atmosphere of earlier techno is frequently perceived as tinted with nostalgia. Today, an audience's sense of 1980s Detroit is now mediated by music, and the music can, in turn, conjure a (fictionalised version of) that place.

Contrary to the suggestion that they are unimaginative works involving sticking a beat on top of a largely already-complete work, techno remixes of minimalist classical pieces are highly diverse: some are remixes of specific pieces in which the original is clearly discernable, while others take a broader inspiration from several works of the artist, or transform the material so extensively that it is barely recognisable. I have shown that – despite the attempts of defenders of classical recomposition to demonstrate its superiority – when it comes to rewritings, there are few practices which are clearly the sole preserve of classical practitioners. Instead, associations retained with specific places contribute to the sense of a piece's origins, sources, and influences. In both classical and techno reworkings, the friction of transplanted elements against each other gives rise to a complex construction of place in music, one which does correspond directly to any actual place. Rather, there are usually traces of multiple places implicit in the remix, which can include among others: 1960s Manhattan, Detroit, the acoustic space in which the audio material was made, and a hint of Europe – and the other place, the virtual place to which the audience escapes.

Notes

1. This catalogue of sounds is listed on the website of the publishers, Boosey and Hawkes. Sirens and alarms were of course one of the signature sounds of minimal techno as well, as Richard Pope indicates (Pope 2011).
2. La Monte Young (b. 1935) is often described as a minimalist. His music frequently involves tones of very long duration.
3. 'Techno producers sample, emulate, and remix the music of minimalist masters.'
4. In both classical and popular spheres, minimal music distinguished itself from other music by eschewing conventional forms and structures – such as climactic points, reprises, or choruses. Instead, there would be repetition, with some kind of variation (changes in texture, phase, addition and removal of small amounts of musical material, subtle transformations).
5. Although the Motown music which had been produced in Detroit during the successful years of the automotive industry has been said to embody certain aspects of industrial production, the sounds of German industrial music were at this time less familiar.
6. Minimalism in art has been described as being characterised by 'an impersonal austerity, plain geometric configurations and industrially processed materials' (Grove Art Online). Minimalism in music and in the visual arts share features including economy of means and the rejection of traditional expressivity.
7. Theories of intertextuality can provide a model for the ways in which the listener's reception of musical material is shaped by their shared knowledge of the previous contexts of its appearance (or the appearance of analogous material) in other work.
8. In fact, back in 1968, Terry Riley created a remix of *You're No Good* by the Harvey Averne Dozen, using two tape machines, which might be comparable. In it, he played on the already-repetitive elements of the song to enable it to transition seamlessly from normal playing into looping.
9. The German group Kraftwerk have been widely acknowledged as one of the early influences in the development of Detroit techno.
10. A form of electronic composition which is based on the arrangement and transformation of recorded sounds, rather than the generation of new sounds through electronic means.
11. Although not falling within either the 'classical minimalism' or 'minimal techno' category, the music of Radiohead also exhibits (to varying degrees on different records) minimal tendencies, and the solo work of band members has involved some exploration of minimalist directions.
12. There are two previous examples, Proverb, which is based on a piece by Perotin, and his version of Sondheim's Finishing the Hat (Chadburn 2012)
13. In a section of documentary film about Richter, the composer is shown marking up a blank score of the Four Seasons in the manner of a conductor.
14. The Kondakion is one of three intermezzo pieces which appear between the 'odes' of the Eastern Orthodox text.
15. Samuel Johnson described this phenomenon using the phrase 'fame is a shuttlecock' – i.e. the continuing discussion and debate (in whatever medium) concerning the material is the only way in which it remains relevant or 'in the air' (I am indebted to Grace Egan for bringing this useful metaphor to my attention).

Works cited

Ankeny, Jason. 'Artist biography of Terry Riley', http://www.legacyrecordings.com/a/#/artist/terry-riley/1137/ (accessed 03 September 2014).

Chamberlin, Daniel (2003). http://www.miaminewtimes.com/2003-09-18/music/party-arty/ (accessed 03 September 2014).

Che, Deborah (2009). Chapter 15, 'Techno: Music and entrepreneurship in post-Fordist Detroit', in Ola Johansson and Thomas L. Bell (eds), *Sound, Society and the Geography of Popular Music*, p. 267.

Cox, Christoph (1999). 'Remix and match', http://www.thefreelibrary.com/Remix+and+match.-a054169953 (accessed 03 September 2014).

Cox, Christoph and Warner, Daniel (2004). Introduction to *Audio Culture: Readings in Modern Music* (New York: Continuum).

Eeles, Cameron (2006). 'Jeff Mills-Blue potential', http://www.residentadvisor.net/review-view.aspx?id=3883 (accessed 03 September 2014).

Fink, Robert (2002). 'Elvis everywhere: Musicology and popular music studies at the twilight of the canon', in Roger Beebe, Denise Fulbrook and Ben Saunders (eds), *Rock over the Edge: Transformations in Popular Music Culture* (Durham, NC: Duke University Press).

Gann, Kyle (2006, 01 January). 'Totalism as a new rhythmic paradigm', *Arts Journal*, http://www.artsjournal.com/postclassic/2006/01/totalism_as_a_new_rhythmic_par.html (accessed 02 October 2014).

Hewett, Ivan (2012, 31 October). 'Vivaldi remixed: Classical music reinvents itself, the telegraph', http://www.telegraph.co.uk/culture/music/classicalmusic/9634354/Vivaldi-remixed-classical-music-reinvents-itself.html (accessed 01 June 2013).

Klein, Joshua and Steve Reich (2006). http://pitchfork.com/features/interviews/6490-steve-reich/ (accessed 03 September 2014).

Large, Pete. Interview with Brandt Bauer Frick, *IDOL Magazine*. http://idolmag.co.uk/music-interview/brandt-brauer-frick (accessed 03 September 2014).

Lawrence, Tim (2009). *Hold onto Your Dreams: Arthur Russell and the Downtown Music Scene, 1973–1992*, First Edition (Duke University Press Books).

Naylor, Tony (2008). 'The strange, lingering death of minimal techno'. http://www.guardian.co.uk/music/musicblog/2008/oct/07/death.minimal.techno.house (Accessed 02 September 14).

Oteri, Carl (2010). 'A conversation with Robert Carl, author of Terry Riley's In C', http://www.newmusicbox.org/articles/A-Conversation-with-Robert-Carl-author-of-Terry-Rileys-In-C/ (accessed 03 September 2014).

Pitera, Dan (July 2010). 'Detroit: Syncopating an urban landscape,' *Places Journal*. https://placesjournal.org/article/detroit-syncopating-an-urban-landscape/ (accessed 19 October 2014).

Pope, Richard (2011). 'Hooked on an affect: Detroit techno and dystopian digital culture', *Dancecult: Journal of Electronic Dance Music Culture* 2(1), pp. 24–44.

Raven, Ben (2008). 'Magda: Minimal techno is mainstream', http://news.beatport.com/blog/2008/08/19/minimal-techno-is-mainstream/ (accessed 03 September 2014).

Reynolds, Simon (2008). *Energy Flash: A Journey Through Rave Music and Dance Culture*, Picador.

Ross, Alex [Untitled] (2006) http://www.stevereich.com/articles/Alex_Ross.html (accessed 03 September 2014).

Rutherford-Johnson, Tim (2013, 27 March). 'The influence engine: Steve Reich and pop music', *New Music Box*, http://www.newmusicbox.org/articles/the-influence-engine-steve-reich-and-pop-music/ (accessed 01 June 13).
Sande, Kiran (2010). 'A brief history of minimalism', http://www.factmag.com/2010/02/01/a-brief-history-of-minimalism/2/ (accessed 03 September 2014).
Savage, Jon (1993). 'Machine soul: A history of techno,' http://music.hyperreal.org/library/machine_soul.html (accessed 03 September 2014).
Sherburne, Philip (2004). 'Digital discipline: Minimalism in house and techno', in Christoph Cox and Daniel Warner (eds), *Audio Culture: Readings in Modern Music* (New York: Continuum).
Sicko, Dan (2010). Techno Rebels: The Renegades of Electronic Funk. (Detroit: Wayne State University Press).
Stolman, Elissa (2012).'Q&A: Robert Hood on techno, minimalism and reviving Detroit', http://www.cmj.com/feature/qa-robert-hood/ (accessed 03 September 2014).
Strickland, Edward (1993). *Minimalism: Origins* (Bloomington: Indiana University Press).
Tingen, Paul. Steve Reich: City Life (Interview (1995). http://www.soundonsound.com/sos/1995_articles/aug95/stevereich.html (accessed 03 September 2014)
Toland, Justin (2011). 'Villalobos/Loderbauer Re: ECM', *Factmag* http://www.factmag.com/2011/07/11/ricardo-villalobos-max-lodebauer-re-ecm/ (accessed 03 September 2014).
Vecchiola, C. (2011). 'Submerge in Detroit: Techno's creative response to urban crisis', *Journal of American Studies* 45(1), pp. 95–111.
Walters, John L (2011). 'Brandt Bauer frick say you can get (slightly) Jiggy with it', http://www.guardian.co.uk/music/2011/aug/04/brandt-brauer-frick-interview (accessed 03 September 2014).
Want, Christopher. Minimalism, *Grove Art Online. Oxford Art Online*. Oxford University Press http://www.oxfordartonline.com/subscriber/article/grove/art/T058397 (accessed 26 October 2014).

Concert and publisher listings

Chadburn, Leo (2012). Music for Umpteen Musicians: Steve Reich Interviewed http://thequietus.com/articles/09261-steve-reich-interview-2 (accessed 03 September 2014)
Listing on the ECM Records website http://player.ecmrecords.com/villalobos_loderbauer/cms/project (accessed 03 September 2014)
Reich Remixed, listing on Nonesuch record label website http://www.nonesuch.com/albums/reich-remixed (accessed 03 September 2014)
Southbank Centre events listing http://ether.southbankcentre.co.uk/5/events/the-brandt-brauer-frick-ensemble/ (accessed 02 September 2014, no longer accessible)
City Life, publisher listing http://www.boosey.com/cr/music/Steve-Reich-City-Life/5922 (accessed 03 September 2014)

DVDs and documentaries

Blue Potential (Live with Montpellier Philharmonic Orchestra) (CD and DVD-V), Satélite K SATKCD053, 2005

Detroit Techno City, 2001. http://youtu.be/fsSMnnJ_YgI (accessed 26 October 2014)
Detroit, the Blueprint of Techno, MuchMusic, 1998
Hi-Tech Soul: The Creation of Techno Music, dir. Gary Bredow, Plexi Film, 2006
Journeys with Max Richter, dir. Yulia Mahr (exerpt from an ongoing project) http://youtu.be/pggJlsbmhtQ (accessed 26 October 2014)
Music is a Place (Interview with Philip Glass), Brian Ritchie https://www.youtube.com/watch?v=fSoRty8LmN4 (accessed 03 September 2014)
Universal Techno, Arte, 1996

Part II
Music, Place and Tourism

6
Abbey Road Studios, the Tourist, and Beatles Heritage

Peter Atkinson

EMI Studios, Abbey Road, London NW8 was renamed Abbey Road Studios by EMI in 1970 after it was made iconic by the eponymous Beatles long-playing record (LP) with its famous cover of the four group members on the zebra crossing before the building. In this chapter I use Urry and Larsen's concept of 'the tourist gaze' to discuss the changing meaning of Abbey Road Studios over the course of several decades and a wider relationship between the work of the Beatles and tourism. I will examine the processes through which the development of the music the group recorded at the studio in the period 1965–1969 became symbolic of wider social and cultural change and argue that some of the songs prefigure the leisure and tourism activities and structures of later decades. In relation to its promotion as an object of the tourist gaze, I note that EMI and the Beatles' company Apple exploited Abbey Road Studios in the 1980s during a period when there was an increased emphasis on national heritage. I conclude by arguing that the coinciding of an aggressive rebranding of 'The Beatles' in the mid-1990s with the retro-aesthetic of the newly named 'Britpop', reinforced the notion of a British rock tradition and lineage. This is seen to elevate Abbey Road Studios into a cultural symbol, giving it further appeal as a tourist destination.

An object of the tourist gaze

Abbey Road Studios is the object of a number of different forms of tourism: Beatles fans attend it; Beatles 'pilgrims' attend it as a 'shrine' (see McCarron 2006: 171–180 for an engaging analysis of the distinction between Beatles pilgrims, fans, and tourists); tourists have themselves photographed on the zebra crossing in front of the studio (and Abbey

Road Studios' own 'Crossing Cam' facility records and archives shots of individuals who cross it – www.abbeyroad.com/Crossing); guided rock tours of London visit the site; the history of the studio is documented in several publications, some of which feature an extensive range of photographs and copies of archive documents (see Lawrence 2012; Lewisohn 1988; Southall et al. 2002); and fans and scholars access and scrutinise the extensive listings, details, and analysis of Beatles recordings made there as provided in an array of literature and websites devoted, or related, to the topic. It should be noted that there are few opportunities for tourists to enter the studios themselves as these are still active and in considerable demand. Abbey Road Studios has provided some special on-site events for visitors but, mostly, tourists visit the zebra crossing outside the studio and view the studio complex from outside its walled and gated perimeters. Some tourists leave their mark with graffiti on the perimeter wall, '[p]ilgrims to the wall' – McCarron calls these, or on the street sign for Abbey Road (McCarron 2006: 175).

For Urry and Larsen pleasure in tourism is derived from making a 'departure', a 'limited breaking' with the everyday and mundane (Urry and Larsen 2011: 3). They note the development of photography in the objectifying activity of 'the tourist gaze', a metaphor – emphasising the visual – for the increasing consumption of sites (sights) as tourism became a popular activity in the mid-to-the-late nineteenth century. The tourist gaze is 'not a matter of individual psychology', but a 'socially patterned' way of seeing. People gaze upon the world 'through a particular filter of ideas, skills, desires and expectations, framed by social class, gender, nationality, age and education' (ibid.: 2). It is constructed through signs derived from many sources, predominantly through media in the modern age, and tourism involves 'the collection of signs' (ibid.: 4). Urry and Larsen note the changing nature of the tourist gaze in the era of globalisation and paraphrase Bauman who defines such conditions as a move 'from a solid, fixed modernity' to a 'liquid modernity'. In the latter, the 'time-space compression' evident since the late 1990s involves not only 'rapid flows of travellers and tourists' but, also, 'complex intersections' between increased 'corporeal travel' and 'virtual and imaginative' travel enabled by digital communications technology (Urry and Larsen 2011: 23). One consequence of globalisation is that 'different countries, or different places within a country, come to specialise in providing particular types of objects to be gazed upon' and Britain 'came to specialize in history and heritage' (ibid.: 55). The Beatles made EMI Studios, Abbey Road, famous in what Samuel terms the 'modernizing hour of the 1960s' (Samuel 1996: 154). This was during a time which Bauman labels 'solid' modernity. For him this was an *'era of mutual engagement*: between the supervisors and the supervised, capital and labour, leaders and their followers, armies at war' (his italic) (Bauman 2000: 11). The methods of production used to make the Beatles' records at Abbey Road Studios reflect such an arrangement of labour. However, the experimentation evident in their later studio productions, the soundscapes and lyrics of which often combine an array of diverse cultural references and images, prefigure the accelerated circulation of images, the global mobility, and the virtual and imaginative travel in liquid modernity. This transition in the group's work is examined below.

First it is useful to recall the early history of EMI Studios as this emphasises the hegemonic associations of its heritage and provides historical context. EMI installed the studio complex in the 'elegant' suburb of St John's Wood in 1931 and it was opened with a rendition of *Land of Hope and Glory* by the London Symphony Orchestra conducted by the celebrated English composer Sir Edward Elgar (Southall et al. 2002: 19). Two blocks away from the famous Lord's cricket ground, where the England cricket team plays its test matches, and around half a mile from Regent's Park (a Royal park, appropriated by Henry VIII), this was an upper middle-class area in the 1960s and, in the modern day, is an extremely desirable and expensive area. The studio is thus associated with the upper-class and what is termed 'the establishment', qualities characterised by the man who was chairman at the time the Beatles were with the label. Sir Joseph Lockwood was an establishment figure, an industrialist and businessman who had held a position of responsibility ensuring food supplies to Europe as countries were liberated towards the end of the Second World War. He was later made a Director of the National Research Development Association before taking over as Chairman of EMI in 1954 where he turned around the company's fortunes and was subsequently made a Knight Bachelor in the New Year's Honours List in 1960. EMI Studios was the 'workshop' of the lower class, provincial Beatles from 1962 until 1970 after being signed to EMI's Parlophone Records label following an audition at Abbey Road in 1962 (Southall et al. 2002: 68) (Figures 6.1).

The Beatles' work at the studios is notable because the way they recorded their music there, and the content of that music, changed dramatically in a short space of time. Considerable critical attention is paid to the difference between the method of recording their first LP *Please Please Me* (Parlophone, 1963) and the methods used for their revered work *Sgt. Pepper's Lonely Hearts Club Band* (Parlophone, 1967) (Everett 1999, 2001; Hertsgaard 1995; Julien 2008a; Womack 2007). For the former, ten songs were recorded at the studio in a continuous

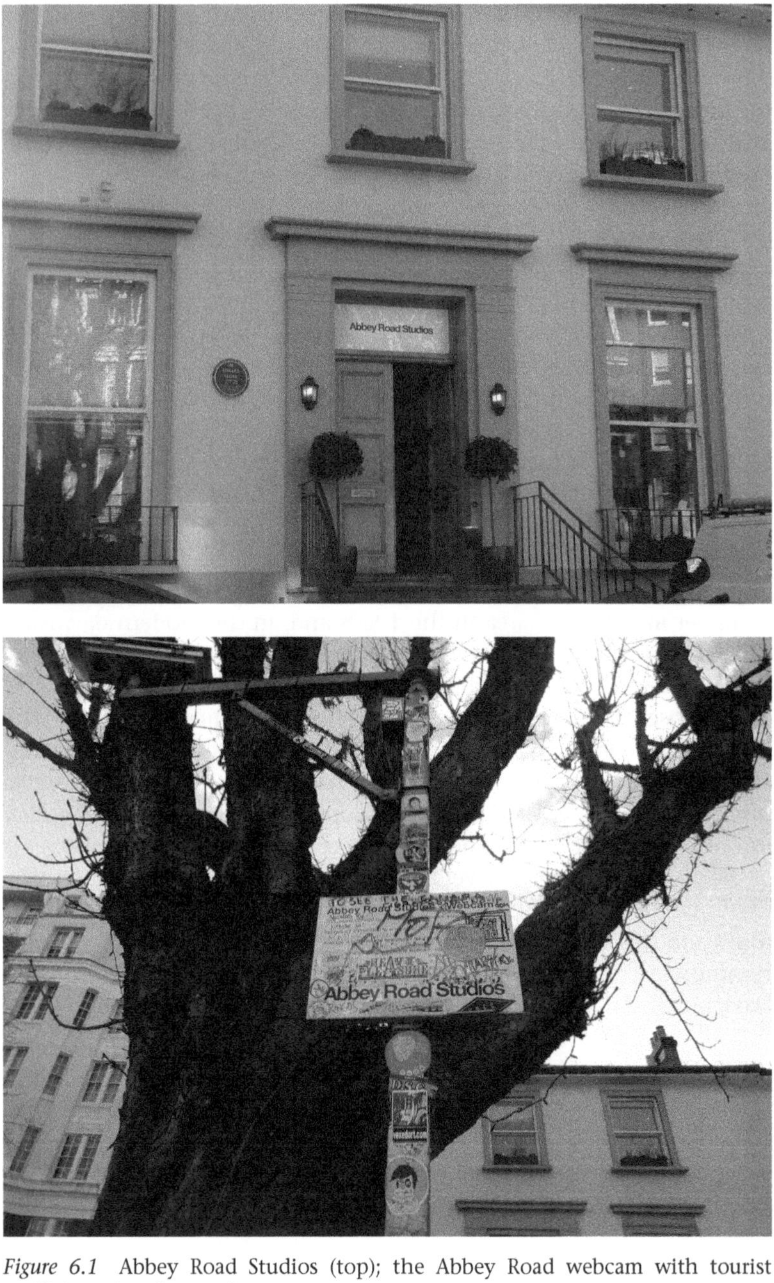

Figure 6.1 Abbey Road Studios (top); the Abbey Road webcam with tourist graffiti on sign (bottom)

Source: Photographs by Kamila Rymajdo.

12 hour session in one day, on two-track recording equipment (EMI had four-track, but considered the Beatles too 'lowbrow' to be afforded access to it, Zolten observes) (Zolten 2009: 39). Essentially, the live sound of the band is recorded at this session, with a little enhancement through over-dubbing on the limited two-track equipment. By contrast, *Sgt. Pepper* took six months to make, involving 700 hours of recording, at a cost of over £25,000 (Julien 2008a: 5). This historic leap in the methods of music production was the result of the Beatles being 'rewarded with the virtual freedom of the studios' because of their commercial success (Southall et al. 2002: 76). From 1965, they worked there on a succession of 'studio-savvy albums' (Zolten 2009: 34). These include: *Rubber Soul* (1965), *Revolver* (1966), the aforementioned *Sgt. Pepper, The Beatles* (1968) (more commonly known as 'the White Album'), and *Abbey Road* (1969). This body of work is widely considered to have changed the face of popular music: Macdonald writes that their recordings comprise 'an outstanding repository of popular art' and a 'cultural document of permanent significance' (Macdonald 1995: 33). The progression of the Beatles' music during this time involved an increased creativity which was realised within the workspace of EMI Studios. From the time of *I Want to Hold Your Hand* (November 1963) the Beatles increasingly explored the possibilities of a recording studio, realising that 'recording no longer had to take place in real time', and that recorded work could be *composed* (Zolten 2009: 40). The paternalistic producer at the studio, George Martin, later Sir George Martin, helped the group fulfil their creative vision, orchestrating skilled productions and 'fashioning' their sound (Womack and Davis 2006a: 2). A variety of technical devices and tricks (overlays, backwards tracking of instruments, phasing, flanging, veri-speed, Artificial Double Tracking (ADT), tape-loops) was used in the sequence of ground-breaking albums. 'Studio technique abounded, and functioned as an instrument itself, wholly integrated into the art of the music', Zolten contends (Zolten 2009: 45). An expanded range of musical instrumentation was also employed (brass, strings, full orchestra, Mellotron, sitar and many more) in order to realise the Beatles' musical vision. Songs simultaneously acquired a new 'literary' dimension (see Burns 2009: 225). The emergence of a new kind of pop song in the early 1960s, profiled by the likes of Bob Dylan, Leonard Cohen, and Joan Baez, 'focused attention on the words, encouraging listeners to assess song as text', Cook and Mercer observe (Cook and Mercer 2000: 97). This is relevant to the concept of the tourist gaze, because a broadening of subject-matter in Beatles songs and the representation of more active and socially mobile subject positions (evident in the

songs of other artists also), anticipated the expansion and increasing diversification of tourist activity in the mid-to-the-late 1960s.

Representative of an expanding experiential reach

To provide some examples of this shift in the subject-matter of Beatles songs and in the music, Cook and Mercer observe a 'sparseness of situational detail and precise reference' in the lyrics of Beatles' earlier work where songs are 'conversation-like' and tend to be about emotional issues (Cook and Mercer 2000: 93). Their narrow, generalised focus – as illustrated in songs such as *From Me to You* (April 1963) and *I Want to Hold Your Hand* – is predictable in music which has its origins in the pop and rhythm and blues genres, their simplicity perhaps necessary for a working band performing regularly for excitable young audiences (ibid.: 96). Later songs 'are far more specific about people, places, ... times' (ibid.: 93). In what MacFarlane terms the 'middle period' in the Beatles' career (1965–1966) the subjects represented broaden and there is a representative gaze at much more detailed and varied social circumstances (MacFarlane 2008: 34). This mirrors the expanding 'experiential reach' in modernity that Wang refers to (Wang 2000: 194). Elsewhere in this volume Mazierska refers to the concision of the form of popular songs; their brief gaze at some aspect or topic is 'touristy'. Songs may be like short films in their depiction of situations (indeed George Martin compares the studio production of the Beatles' later work to film editing, as Kimsey notes) (Kimsey 2008: 134). Davis writes that the Beatles 'metamorphosed, discarding childish, hackneyed, borrowed forms to blossom into the most admired, most studied, ... songwriters of our age' (Davis 2014: 28). I argue that the change implies an increased 'social mobility', to use Marwick's term, which refers to the changes in lifestyle, socio-cultural awareness and expectations that derive from increased wealth and social status (Marwick 2003: 127). Thus, in the sexist song *Norwegian Wood (This Bird has Flown)* (December 1965) the songwriters portray life among London's wealthy young elite. The girl who is the object of the protagonist's thwarted desire has indulged in the importation of Scandinavian design, which was fashionable at the time. The 'Norwegian wood' in the song acts as a metaphor for the 'counterfeit' nature of the relationship portrayed, which is conducted in a newly decorated flat: the wood the couple refer to is in reality 'cheap pine', Decker observes (Decker 2009: 79; see also Miles 1997: 270). The song is indicative of an expanded experiential reach as it incorporates concepts such as female assertiveness and independence (she rejects his advances – in her own flat – because she has work 'in the morning'), fashions in contemporary interior design, and male violence in a middle-class social setting (as the insinuation in the lyric is that the protagonist sets the wood interior alight the next morning in revenge for his rejection) (Miles 1997: 270). MacDonald comments that the song's 'unusual instrumental colour' is provided by a sitar (played by George Harrison) doubling the descending main melody line (MacDonald 1995: 132). He further notes that Harrison had been interested in the sitar since he had heard it being used to 'spice up' the soundtrack of the Beatles movie *Help!* (UK, 1965). As one of the authors of *Norwegian Wood*, Lennon was introduced to the instrument through hearing, along with the other Beatles, 'exotic' *raga* phrases played to him by David Crosby of the American group the Byrds in August 1965 when the Beatles rented a house in Los Angeles during their American tour. MacDonald contends that elements of the song are influenced by the drone common in all Indian classical music and that the descending melody is an attempt at reproducing Oriental intervals, which Lennon had also been introduced to by the Byrds (MacDonald 1995: 132). These influences, and the way in which the Beatles were introduced to them, are indicative of broadening social and cultural horizons associated with increased mobility and anticipate the increased global communication, mass travel and tourism, and the lifestyle consumption which have developed since then. *Day Tripper* (December 1965) refers more directly to issues of leisure and tourism although the punning reference to the 'tripper', an occasional drug-user, as a 'Sunday driver' is a metaphor (McCartney quoted in Miles 1997: 209). Nonetheless, the subject of the song plainly exercises a range of lifestyle choices. Lennon and McCartney were the main writers in the Beatles and they usually conferred in the creation of songs. As working-class, provincial boys from Liverpool, their new experiences, both in London where they mingled with the creative elite, and abroad as they travelled, informed their song-writing. The portrayals within their middle and late period songs may be said to reflect the 'wide-ranging structural and cultural developments within contemporary societies' (Urry and Larsen 2011: 97). Urry and Larsen observe that the 'rapid and significant change' in tourist practices cannot be separated from such developments,

> [M]ass communications have transformed the tourist gaze which is increasingly bound up with, and is partly indistinguishable from, all sorts of other social and cultural practices. This has the effect, as 'tourism' *per se* declines in specificity, of generalising the tourist

> gaze – people are much of the time 'tourists' whether they like it or know it. The tourist gaze is intrinsically part of contemporary experience.... (Urry and Larsen 2011: 97)

That the Beatles and their work are emblematic of the 1960s zeitgeist is precisely what has made them object of the modern tourist gaze (see Whiteley 2009). As a result of this, Abbey Road Studios, which was the production base for the main output of the Beatles' music, becomes a symbol for the creative intervention made by the group in their time. As Urry and Larsen note, paraphrasing MacCannell, all tourists 'embody a quest for authenticity', this quest being 'a modern version for the human concern with the sacred' (Urry and Larsen 2011: 10). Abbey Road Studios is the authentic article, the place of the Beatles' work, and it is the object of tourist activity.

The 'late period' Beatles (1967–1969) finds the horizons of the representative gaze in their songs further extended (MacFarlane 2008: 35). This reflected their experience of travelling extensively for purposes of work, leisure, and self-fulfilment: *Child of Nature* (1968, never released) is about their experiences in Rishikesh as guests of guru Maharishi Mahesh Yogi; *Sexy Sadie* (November 1968) was originally titled *Maharishi* and is a polemic about the yogi's alleged hypocrisy during their visit; *Back in the USSR* (November 1968) begins with the sound of a Boeing jet taking off and 'satirizes the absurdities of the Cold War by situating nostalgia within Soviet Georgia', Whiteley asserts (Whiteley 2009: 214). There is a depiction of fantasy, spectacle, and surreality influenced by the fantasy worlds depicted in English literature, particularly those of Lewis Carroll, in *Lucy in the Sky with Diamonds* (June 1967) and *I Am the Walrus* (November 1967) (MacDonald 1995: 190; Zolten 2009: 51). *Being for the Benefit of Mr Kite* (June 1967) depicts a Victorian circus of 1893 (being inspired by a poster of the event Lennon purchased in an antique shop). *A Day in the Life* (June 1967) is a dystopian evocation of fate and death in modern day life. The *Magical Mystery Tour* concept, which included an EP release and television film (December 1967), is meanwhile a psychedelic-flavoured, carnivalesque inversion of the staid, British working-class holidaying traditions of the industrial age. (This was partly inspired by Ken Kesey's Merry Pranksters' road trip around America promoting hallucinogenic drugs in a multi-coloured bus in 1964.) (Womack 2007: 191). Again, such subject-matter anticipates the range of gazes and activities of the 'new tourism' in the age of post-Fordist consumption. This is 'segmented, flexible and customized' and a host of different, often themed, tourist experiences are offered to the consumer and might include literary tourism, dark tourism, heritage sites and theme parks (Urry and Larsen paraphrasing Poon 2011: 52–53). In their revered 1967 album the Beatles create a band – 'Sgt. Pepper's Lonely Hearts Club Band' – that is a simulation. The album's opening 'sound collage', of crowd noises and the band tuning up, are sourced from an existing Abbey Road stock of recordings and sound effects and used at the whim of the Beatles. This technical effect acts as 'narrative source and as a psychological way' of creating a mood of 'shared festivity', Whiteley suggests (Whiteley 1992: 40). The concept of the work mirrors the simulated structure of 'themed' tourist spaces and attractions – as the line 'a splendid time is guaranteed for all' from the album track *Being for the Benefit of Mr Kite* suggests. Such themed attractions represent, for Urry and Larsen, 'a paradoxical mix of presence and absence, here and there', and are also 'typified by high capital investment' – just like with *Sgt. Pepper* (Urry and Larsen 2011: 125).

In songs, 'sound-shape, together with the socio-cultural element superimposed upon it, consolidate to form a distinct form of communication', Whiteley asserts. The musical landscapes and lyrical worlds of this Beatles work, and that of other musicians at that time, broaden the *scope* of representation in popular song (Whiteley 1992: 3). The band was a primary influence in the development of modern rock/pop music which was considered to be a vehicle of expression in the later 1960s 'counter-culture' when it was 'thought to say things of cultural and political significance, to have a message' (ibid.: 1). Such music has an overtly ideological dimension and therefore impacts upon the 'particular filter of ideas, ... desires and expectations' through which the tourist gaze is 'socio-culturally framed' (Urry and Larsen 2011: 2). The most disturbing example of the influence of rock music of the mid-to-the-late 1960s is that of Charles Manson who thought the Beatles were 'directing secret messages to him via their songs' – and through *Helter Skelter* (November 1968) in particular. Manson blamed 'the White Album' for inciting the murder of seven people by members of his 'Family' in 1969 in California (Miles 1997: 488). There are currently websites that promote tours of sites relating to the murders, thus illustrating how cultural interpretation of rock songs translates into tourist productions.

Abbey Road and Beatles heritage

Having been the workshop in which Beatles' creativity was crafted into an internationally significant and culturally valued product, Abbey Road Studios provided the stage for an important event in the development of globalisation in 1967. The significance of the Beatles as emblems of nation and of the so-called 'cultural revolution' of the 1960s was reflected in their being chosen to represent Britain in the first global satellite television link-up, *Our World*. This was broadcast by the BBC to an estimated audience of 400 million in June that year with the band singing *All You Need is Love* (1967) from Abbey Road Studio One, produced by George Martin. This is an early example of the construction of an image whereby the Beatles, in their workplace of Abbey Road, are selected to be representative of the nation on the global stage.

In the 1980s the connotations of Abbey Road Studios changed. It started to be seen as part of English heritage. I now outline how this aspect of the studio was emphasised at this time. Since the National Heritage Act of 1983, and the increasing privatisation of the heritage/museum 'industry', cultural observers have noted the importance of heritage to tourism (Hewison 1987; Marwick 2003: 322–323; Samuel 1996). Focusing on it 'affects both what overseas visitors expect to gaze upon and what attracts British residents to spend time holidaymaking within Britain' (Urry and Larsen 2011: 55). There was some feeling at the time that this was a reaction to de-industrialisation of the nation, and Urry and Larsen suggest that '[h]eritage is seen as involving a strong sense of lineage and inheritance' and it has 'an identity-conferring status' (ibid.: 141). Samuel writes that 'perhaps as a result of the collapse of ideas of national destiny', there is the increasing importance of 'memory places' and in 'ideas of the national past'. Certain locations do the 'memory work which in earlier times might have been performed by territorial belonging' (Samuel 1996: 39). As Waterton observes, the places that are promoted as being representative of the nation are mostly those favoured as one of the 'cultural symbols of an elite group' – the white middle class. Such touristic places 'become ideological spaces', she concludes (Waterton 2010: 155).

The Beatles disbanded in 1970. John Lennon was murdered in 1980 and, from the early 1980s, EMI increasingly realised the commercial and promotional potential of their holdings of archival Beatles material and their ownership of Abbey Road Studios. These were employed to attract tourists, and to create new Beatles products, in summer 1983 when Studio Two opened to visitors in the event *The Beatles at Abbey Road Studios*, initiated by Ken Townsend, studio manager, who had worked with the group (Southall et al. 2002: 169). Again, stressing the value of authenticity to the tourist experience, part of the studio was set up 'exactly as it always had been for Beatles recordings, compete with all the original microphones', for the event which was a considerable success, attracting 22,000 visitors. A two-part video including rare footage, photos, interviews and previously unreleased versions of songs was also released (ibid.: 170).

New compact disc (CD) technology provided further opportunity to exploit Beatles heritage. A CD version of *Sgt. Pepper* was released in June 1987 on the twentieth anniversary of the original, and the release was celebrated by a party, again in Abbey Road Studio Two, the site of the original recording. In the same year, Mark Lewisohn was commissioned to write a book detailing all the Beatles' recording sessions. *The Complete Beatles Recording Sessions: The Official Story of the Abbey Road Years* was published in 1988, a lavish publication which featured collages of text, photographs, record covers and reproductions of studio recording sheets and other studio documents. This is a fetishisation of place, artefacts and events as heritage, and enhances the appeal of Abbey Road Studios to the tourist. In 1992 *The Making of Sgt. Pepper*, a television documentary featuring producer George Martin and other members of the Beatles, was broadcast on Independent Television's (ITV) *The South Bank Show* to commemorate the 25th anniversary of the record's release. Meanwhile the Beatles themselves recognised the value of archival sound recordings as heritage product in the CD age and were planning the *Anthology* project.

The *Anthology* collection of products was designed to rebrand the Beatles some 30 years after they had initially achieved near global recognition. Decker argues that in 1991 the Beatles' 'reputation for cool ebbed low' among teenagers increasingly fed on a diet of MTV audio-visual pop (Decker 2006: 183). He suggests that McCartney in particular was interested in 'protecting the Beatles' textual legacy' and spearheaded 'a massive revision of Beatles history' (ibid.: 185). This involved a series of archival releases including the CDs *Live at the BBC* (1994), *Anthology 1* (1995a), *Anthology 2* (1996a), *Anthology 3* (1996b), *The Beatles Anthology* television documentary and video cassette (1996c) (later released as a DVD boxed set in 2003) and *The Beatles Anthology* book (Roylance 2000). As with Lewisohn's book, *The Complete Beatles Recording Sessions: The Official Story of the Abbey Road Years*, the *Anthology* releases made considerable use of archive material – recordings, documents and the testimony of those involved in the recording of the original music – derived from Abbey Road Studios. A virtual Beatles 'reunion' was also contrived. Two of Lennon's private demo tracks from the 1970s, *Free as a Bird* and *Real Love*, were engineered using studio technology by the remaining Beatles – with the help of producer Jeff Lynne for the former, and George Martin for the latter – into 'new' Beatles tracks and released as singles (Burns 2000: 178). These simulations also appeared as tracks on *Anthology 1* and *2* respectively. The *Anthology* series was an exercise in creating heritage products for consumption and stressing the value of authenticity. The CDs comprise mainly of out-takes from recording sessions through which it is possible to observe the development of individual songs. Combining these with the *Anthology* documentary, the Beatles 'attempted to create the simulacra of Beatlemania', Decker contends. By 'accentuating the band's textual corpus', he argues, 'McCartney established the illusion that the Beatles had yielded their position as subject and were now in a position to reminisce over their considerable output' (Decker 2006: 187). As a consequence of this rebranding the Beatles enjoyed a second 'phoenix-like rise' in the mid-1990s (ibid.: 184).

It is notable that the later *Anthology* CD releases in 1996 garner a more favourable critical response. This suggests that it took time for the value of such documentary works to be appreciated. In the *RollingStone* review of *Anthology 2* of 4 April 1996, Jerry McCulley writes that the 44 tracks from the period February 1965 to February 1968 'document what is arguably the most creative 36 months for one band in rock history'. Noting that the majority of the album consists of early takes of songs, he admires the *Strawberry Fields Forever* selection of out-takes which 'quantum-leaps in three cuts from a troublesome acoustic demo' to 'a technically problematic masterpiece'. He concludes that such tracks 'offer a compelling human story along with one of the most crystalline definitions of *synergy* in popular culture' (McCulley 1996). In the *RollingStone* review of the Beatles *Anthology 3*, Puterbaugh observes that, for more than two decades after the Beatles broke up, the band and their producer 'insisted that everything of quality that they created in the studio was already a matter of record'. However, he contests that,

> The extraordinary 1968 demos at the start of *Anthology 3* – seven songs taped in gorgeous, unplugged form at George Harrison's home in Esher, England, shortly before the sessions for the epic double album *The Beatles* ... – show that those who make history are often the least qualified to judge it. ... This is warm, intimate music making, a rare close-up of the Beatles in private, creative ferment, and it is one of the many reasons why the three double CDs' worth of rough cuts and outtakes in the *Anthology* series ultimately enhance rather than dilute the legacy and wonder of the Beatles. (Puterbaugh 1996)

The release of these raised the profile of Abbey Road Studios, the products seeming to bring to life the process of the Beatles recording there. Its heritage is emphasised as the scope of knowledge of the Beatles' working methods is extended with the repackaging of material from the studio's archives in new formats delivered through a diversity of media forms (audio media (CD), television, audio-visual recordings (video and DVD) and publishing). As noted earlier, mass communication technologies and the 'information society', means that the tourist gaze in contemporary society is increasingly less distinct from 'other social activities' (Urry and Larsen 2011: 97). Urry and Larsen write that,

> [C]ulture as an economy of signs is more central in the organisation of present-day societies. There has been a dissolving of the boundaries, not only between high and low cultures, but also between different cultural forms, such as tourism, art, education, photography, television, music, sport, shopping and architecture. (ibid.: 97)

In such a milieu, the *Anthology* collection synergistically made the Beatles and Abbey Road Studios newly visible in the mid-1990s. Utilising Waterton's conceptualisation of the process of imaging in heritage tourism: because it is so strongly associated with the Beatles, Abbey Road Studios, with its iconic zebra crossing image, is 'pushed forward' as part of a 'limited range of apparently consensual images' to define and articulate 'the story of the nation' and its heritage (Waterton 2010: 159). Samuel speaks of 'ministering to an appetite for roots' and such functions may be ascribed to Abbey Road Studios as it becomes focus of the modern tourist gaze (Samuel 1996: 39). The granting of Grade II Listing status confirms this. The Beatles, and the studio, are seen as part of an English music tradition, in a similar way in which literary figures such as Shakespeare, Dickens, and the Lakes poets are seen as part of an English literary tradition. The studio site has become one of the 'idealized images' of '[i]nstitutionalized heritage tourism' in England (Waterton 2010: 158).

Related events in 1995–1996 meanwhile emphasised the studio's heritage as will now be observed. DeNora notes that the past seems to come alive to 'its soundtrack', the *Anthology* releases provide the added value of contextual information to this process (DeNora 2000: 67). Decker validly claims that the Beatles 'carefully manipulated their symbolic valuation to augment the economic value of their canon' with *Anthology* (Decker 2006: 193). It has been seen above that the documentation of the developing creative process enhances the Beatles' reputation as artists of the rock genre in the mid- and late 1960s. The final aspect in the chronology of events which I highlight as being key to the promotion of Abbey Road Studios as a tourist destination, and object of the tourist gaze, occurred in the same year that *Anthology 1* was released and *The Beatles Anthology* television series started. In 1995, the

'1960s retro-aesthetic' of what became termed 'Britpop' was the subject of attention from British media (Bennett and Stratton 2010: 1).

Look back in 1995–1996

Britpop was a generic label for the music of several white English, mainly male, guitar bands – Blur, Oasis, Pulp, Suede, Sleeper and Dodgy who emphasised a 'lineage' from the rock/pop music of the 1960s and the English punk of the 1970s (Du Noyer 2010: 255). Bennett and Stratton write that it was,

> branded by music journalists and critics alike as a critical resurgence of British ... or more specifically, 'English', popular music that rekindled the spirit of the mid-1960s 'British' invasion of the US by groups such as the Beatles, the Rolling Stones, the Kinks, the Who and the Small Faces. Indeed, a number of these groups, notably the Beatles, the Kinks and the Small Faces were frequently cited as key influences by leading Britpop artists such as Blur and Oasis. (Bennett and Stratton 2010: 1)

Partly the result of 'the Movementism beloved by the music papers', *The Guardian* claimed on 8 December 1995, Britpop also gained national exposure as the result of the '[t]abloidized' Blur versus Oasis battle for top spot in the record chart in August 1995, an echo of a (mythological) rivalry between the Beatles and the Rolling Stones in the mid-1960s (O'Hagan 1995: 20). Britpop was seen as a challenge to the hegemony of US popular music, particularly grunge and rap (Bennett and Stratton 2010: 1; Harris 2003: 88; Whiteley 2010: 55). The punning phrase 'Cool Britannia' meanwhile began to be used in 1996, carrying echoes of the self-confident, mid-1960s 'Swinging London' (Huq 2010: 90).

The simultaneous media profiling of Britpop and the rebranding of the Beatles in 1995–1996 crystallises a moment in which an ideological and commercial investment is made in the past. This has generic and ideological implications. The *Anthology* CDs comprise *unfinished* product until now 'not considered worth releasing' – a 'collection of scrappy old Beatles demo tapes, TV recordings, and studio outtakes', Caroline Sullivan argues in *The Guardian* on 24 November 1995 on the release of *Anthology 1*. To relate the *Anthology* set to Britpop: Huq contends that the sub-genre 'began as an offshoot' of the British 'indie' music scene' (Huq 2010: 93). Bannister refers to 'indie guitar rock' that defined itself against its 'other' – 'mainstream', manufactured music – in the 1980s and early 1990s (Bannister 2006: 87). It does so, he argues, by rearticulating 'the traditional rock insistence on the superiority of live performance to recordings' and qualities of 'liveness' in recordings by 'the minimisation of technological mediation' are favoured (ibid.: 92). For indie fans and artists, 1960s jangly guitar-pop is seen as 'pure', authentic. There is a tendency to idealise (a construction of) that epoch with the creation of 'a transcendental, ahistorical ideal', Bannister argues (ibid.: 87). Indie guitar rock looks back towards a selective body of valued, canonical works (particularly by guitar bands), having a 'huge investment in a version of the past, in its own voicing' of 'rock tradition' (ibid.: 81). The *Anthology* set affirmed the Beatles' position within the rock/pop canon in this respect. It documented their development from a raw grassroots band enthusiastically copying American musical styles, the collection of out-takes also exhibiting the unfinished qualities of liveness and authenticity valued by indie guitar rock.

Whiteley views use of influences from the 1960s as 'pop cultural revivalism – the imagery of modern culture as a data base and dressing-up box' (Whiteley 2010: 55). Nonetheless, the indigenous rock/pop heritage is emphasised by Britpop and the simultaneous release of the *Anthology* archive material. This enhances the notion of a British, or English, music tradition and this, in turn, promotes identification with Abbey Road as a landmark in this tradition – hence promoting it as an object of the tourist gaze. As Urry and Larsen observe, discourses of heritage and memory authorise 'different kinds of gaze' in tourism (Urry and Larsen 2011: 19). Whiteley offers that 'it is [...] possible that by selectively invoking the sound and sensibility of English popular culture of earlier eras', such as white male guitar-pop of the 1960s, Britpop 'managed to erase the troubling reminder that Britain is a multi-ethnic society' (Whiteley 2010: 57). However, as noted earlier, the tourist gaze is shaped by, and reflects, dominant ideologies. For Wang, tourism offers a homogenised, simplified, idealised, and beautified spectacle (Wang 2000: 165). The pushing forward of 'consensual images', mentioned above, is at the expense of *others*, not highlighted, not foregrounded in the practices of representation of the dominant group (see Waterton 2010: 155).

It has been seen that the *Anthology* series marketed in the mid-1990s centres on Abbey Road Studios. Connell and Gibson write of '[m]ythologising the local' and of localised music 'scenes' which evolve partly because local music cultures are distinctive, but partly, also, because there is a 'fetishisation of localities' associated with particular 'sounds' (Connell and Gibson 2003: 110). For them,

> The identification of musical difference through regional sounds is an integral component of the fetishisation of place – securing the

> 'authenticity' of local cultural products in particular physical spaces as they move through national and global economies. Connections to place emphasize roots and points of origin. Remaining 'true' to one's roots emphasises credibility. (ibid.: 111)

The release of archive holdings of recorded Beatles studio out-takes at a time when Britpop was attracting much media attention mutually enhances the notion of a British, or English, music tradition. The profiling of authentic practices of music-making in the former, and the retro-aesthetic of the latter, extend the discursive scope relating to English popular music heritage. On the day that Abbey Road was given Grade II Listed status in February 2010, Simon Thurley, Chief Executive of English Heritage, said:

> Some of the most defining sounds of the 20th century were created within the walls of the Abbey Road Studios. English Heritage has long recognised the cultural importance of Abbey Road – it contains, quite simply, the most famous recording studios in the world which act as a modern day monument to the history of recorded sound and music. The listing of the building is a welcome acknowledgement of the contribution the studios have made to our musical heritage.... Listing is a way of saying that a building is special and that every care should be afforded to decisions affecting its future. English Heritage warmly welcome EMI's appreciation for the cultural value embodied in the building and their understanding that listing is an appropriate way to recognise that value. (English Heritage 2010)

Beatles tourism focuses primarily on Liverpool, the focus there is on the *origins* of the Beatles. In the case of Liverpool, the Beatles' 'past has been summoned to rescue the present', using Hewison's assessment of the function of the heritage industry, for the city was in economic decline, even when it was famous for nurturing unique pop music in the 1960s (Hewison 1987: 21; see also Cohen 2007). Interestingly, items featuring Abbey Road iconography (including reproductions of the album cover on cards and coasters, and of the street sign on fridge magnets) are the best sellers in a range of Beatles souvenirs at the Museum of Liverpool gift shop, retail manager Karen O'Connor observes. This is surprising considering that there is also a range of goods on sale which features images of the group's Liverpool past. Abbey Road Studios, in economically dynamic London, is meanwhile part of a vital *present*: it remains a working studio, active in a vibrant popular music scene. The continued, and extensive, critical attention paid to archival Beatles material ensures its discursive buoyancy. It is an active site literally and virtually, the latter being a vital prerequisite of the modern tourist gaze. Once the workshop of the Beatles in the fixity of solid modernity, Abbey Road now hosts discursive engagement which derives from that history in the electronic fluidity of liquid modernity. Authenticity, Connell and Gibson argue, 'hangs on nostalgia' and nostalgia 'can be seen as a new way of imagining communities, harnessed in by the post nation-state, an attempt at a connivance of a recovery of a lost childhood, a return to the m(other) land' (Connell and Gibson 2003: 272). Yesterday is engaged through Abbey Road in the fast flow of global traffic and tourism in liquid modernity. There are places some communities are persuaded to remember.

Works cited

Bannister, Matthew (2006). '"Loaded": Indie guitar rock, canonism, white masculinities', *Popular Music* 25(1), 77–95.
Bauman, Zygmunt (2000). *Liquid Modernity* (Cambridge: Polity Press).
The Beatles (1995a). *The Beatles Anthology 1*, Apple.
The Beatles (1995b). *The Beatles Anthology*, television documentary, Apple for ABC-TV.
The Beatles (1996a). *The Beatles Anthology 2*, Apple.
The Beatles (1996b). *The Beatles Anthology 3*, Apple.
The Beatles (1996c). *The Beatles Anthology*, video cassette, Apple.
Bennett, Andy and Jon Stratton (2010). 'Introduction', in A. Bennett and J. Stratton (eds), *Britpop and the English Music Tradition* (Burlington: Ashgate), pp. 1–7.
Burns, Gary (2000). 'Refab four: Beatles for sale in the age of video', in Ian Inglis (ed.), *The Beatles: Popular Music and Society, A Thousand Voices* (Basingstoke: MacMillan Press Ltd), pp. 176–188.
Burns, Gary (2009). 'Beatles news: product line extensions and the rock canon', in Kenneth Womack (ed.), *The Cambridge Companion to the Beatles* (Cambridge: Cambridge University Press), pp. 217–229.
Cohen, Sara (2007). *Decline, Renewal and the City in Popular Music Culture: Beyond the Beatles* (Burlington: Ashgate).
Connell, John and Chris Gibson (2003). *Sound Tracks: Popular Music, Identity and Place* (London: Routledge).
Cook, Guy and Neil Mercer (2000). 'From me to you: Profligacy in the language of the Beatles', in Ian Inglis (ed.), *The Beatles: Popular Music and Society, A Thousand Voices* (Basingstoke: MacMillan Press Ltd), pp. 86–104.
Davis, Hunter (2014). 'Perfect storm', *New Statesman*, 22–28 August 2014, pp. 26–29.
Decker, James M. (2006). '"Baby you're a rich man": The Beatles, ideology and the cultural moment', in Kenneth Womack and Todd F. Davis (eds), *Reading the Beatles: Cultural Studies, Literary Criticism, and the Fab Four* (Albany: State University of New York Press), pp. 183–194.
Decker, James M. (2009) '*Rubber Soul* and the transformation of pop', in Kenneth Womack (ed.), *The Cambridge Companion to the Beatles* (Cambridge: Cambridge University Press), pp. 75–89.

DeNora, Tia (2000). *Music in Everyday Life* (Cambridge: Cambridge University Press).

Du Noyer, Paul (2010). *In the City: A Celebration of London Music* (London: Virgin Books).

English Heritage (2010). http://www.english-heritage.org.uk (accessed 15 August 2014).

Everett, Walter (1999). *The Beatles as Musicians: Revolver Through the Anthology* (Oxford: Oxford University Press).

Everett, Walter (2001). *The Beatles as Musicians: The Quarry Men Through Rubber Soul* (Oxford: Oxford University Press).

Harris, John (2003). *The Last Party: Britpop, Blair and The Demise of English Rock* (London: Harper Perennial).

Hertsgaard, Mark (1995). *A Day in the Life: The Music and Artistry of the Beatles* (New York: Delacorte Press).

Hewison, Robert (1987). *The Heritage Industry: Britain in a Culture of Decline* (London: Methuen).

Huq, Rupa (2010). 'Labouring the point? The politics of Britpop in "New Britain"', A. Bennett and J. Stratton (eds), *Britpop and the English Music Tradition* (Burlington: Ashgate), pp. 89–102.

Julien, Olivier (2008a). '"Their production will be second to none": An introduction to *Sgt. Pepper*', in Olivier Julien (ed.), *Sgt. Pepper and The Beatles: It Was Forty Years Ago Today* (Aldershot: Ashgate), pp. 1–9.

Julien, Olivier (2008b). '"A lucky man who made the grade": *Sgt. Pepper* and the rise of a phonographic tradition in twentieth-century popular music', in Olivier Julien (ed.), *Sgt. Pepper and The Beatles: It Was Forty Years Ago Today* (Aldershot: Ashgage), pp. 147–169.

Kimsey, John (2008). 'The watchamucallit in the garden: *Sgt. Pepper* and fables of interference', in Olivier Julien (ed.), *Sgt. Pepper and The Beatles: It Was Forty Years Ago Today* (Aldershot: Ashgate), pp. 121–138.

Lawrence, Alistair (2012). *Abbey Road Studio: The Best Studio in the World* (London: Bloomsbury).

Lewisohn, Mark (1988). *The Complete Beatles Recording Sessions: The Official Story of the Abbey Road Years* (London: Hamlyn).

McCarron, Kevin (2006). 'A universal childhood: Tourism, pilgrimage, and the Beatles', in Kenneth Womack and Todd F. Davis (eds), *Reading the Beatles: Cultural Studies, Literary Criticism, and the Fab Four* (Albany: State University of New York Press), pp. 169–181.

McCulley, Jerry (1996), *RollingStone* review of The Beatles *Anthology 2*, http://www.rollingstone.com (accessed 17 August 2014).

MacDonald, Ian (1995). *Revolution in the Head: The Beatles' Records and the Sixties* (London: Pimlico).

MacFarlane, Thomas (2008). 'Sgt. Pepper's quest for extended form', in Olivier Julien (ed.), *Sgt. Pepper and The Beatles: It Was Forty Years Ago Today* (Aldershot: Ashgage), pp. 33–44.

Marwick, Arthur (2003). *British Society Since 1945*, 4th edn. (London: Penguin).

Miles, Barry (1997). *Paul McCartney: Many Years from Now* (London: Secker and Warburg).

O'Hagan, Sean (1995). 'The great white hopes', *The Times*, 2 December, pp. 20–21.

Puterbaugh, Parke (1996). *RollingStone* review of The Beatles *Anthology 3*, http://www.rollingstone.com (accessed 16 August 2014).

Roylance, Bruce (ed.) (2000). *The Beatles Anthology* (London: Cassell/Weidenfeld & Nicolson).
Samuel, Raphael (1996). *Theatres of Memory, Vol.1: Past and Present in Contemporary Culture* (London: Verso).
Southall, Brian, Peter Vince and Allan Rouse (2002). *Abbey Road: The Story of the World's Most Famous Recording Studios* (London: Omnibus Press).
Sullivan, Caroline (1995). *The Guardian*, www.theguardian.com (accessed 16 August 2014).
Urry, John and Jonas Larsen (2011). *The Tourist Gaze 3.0*, 3rd ed. (London: SAGE).
Wang, Ning (2000). *Tourism and Modernity: A Sociological Analysis* (Oxford: Pergamon).
Waterton, Emma (2010). 'Branding the past: The visual imagery of England's heritage', in Emma Waterton and Steve Watson (eds) *Culture, Heritage and Representation: Perspectives on Visuality and the Past* (Farnham: Ashgate Publishing), pp. 155–172.
Whiteley, Sheila (1992). *The Space Between the Notes: Rock and the Counter-Culture* (London: Routledge).
Whiteley, Sheila (2008). '"Tangerine trees and marmalade skies": Cultural agendas or optimistic escapism?', in Olivier Julien (ed.), *Sgt. Pepper and The Beatles: It Was Forty Years Ago Today* (Aldershot: Ashgage), pp. 11–22.
Whiteley, Sheila (2009). 'The Beatles as zeitgeist', in Kenneth Womack (ed.), *The Cambridge Companion to the Beatles* (Cambridge: Cambridge University Press), pp. 201–216.
Whiteley, Sheila (2010). 'Trainspotting: The gendered history of britpop', A. Bennett and J. Stratton (eds), *Britpop and the English Music Tradition* (Burlington: Ashgate), pp. 55–70.
Womack, Kenneth and Todd F. Davis (2006a). 'Dear sir or madam, will you read my book?' in Kenneth Womack and Todd F. Davis (eds), *Reading the Beatles: Cultural Studies, Literary Criticism, and the Fab Four* (Albany: State University of New York Press), pp. 1–6.
Womack, Kenneth and Todd F. Davis (2006b). 'Mythology, remythology and demythology: The Beatles on film', in Kenneth Womack and Todd F. Davis (eds), *Reading the Beatles: Cultural Studies, Literary Criticism, and the Fab Four* (Albany: State University of New York Press), pp. 97–110.
Womack, Kenneth (2007). *Long and Winding Roads: The Evolving Artistry of the Beatles* (London: Continuum).
Zolten, Jerry (2009). 'The Beatles as recording artists', in Kenneth Womack (ed.), *The Cambridge Companion to the Beatles* (Cambridge: Cambridge University Press), pp. 33–61.

7
East Meets West: Tallinn Old Town and Soviet Estonian Pop Music on Screen

Eva Näripea

Due to its geographical position between Russia and Western Europe, Estonian history is one of a so-called 'border state'. Over time, the country has been colonised by both its Western and Eastern neighbours. Nevertheless, its modern cultural identity, initially constructed during the National Awakening in the nineteenth century, is firmly situated in the West, and in particular in the Germanic sphere of influence. When Soviet Russia annexed Estonia and the other Baltic states during the Second World War, an era of enduring cultural conflict began, characterised by a struggle of the locals to resist Russification, and to an extent to come to terms with the inevitable presence of this powerful 'savage coloniser'. From the very beginning, the Baltic countries became a 'Western oasis', the Soviet West, serving as a desirable destination of internal tourism, as well as a showcase of 'progressive' Soviet culture. In particular, the Estonian capital city of Tallinn, and more precisely its well-preserved medieval Old Town, now a UNESCO world heritage site, became an interesting place for negotiations between conflicting ideologies and (national) identities, and an important arena for (re)presentations of power, resistance, and adaptation. Its unique built environment provided a point of gravity to the Soviet Estonian culture, especially in the 1960s and early 1970s when the Old Town and the broader subject of the medieval heritage became extraordinarily topical for both academic circles (mostly in heritage studies of applied nature) and mass culture. In the latter, the increasingly fashionable retro-medieval imagery spread across a wide spectrum of creative enunciations and resulted in peculiar hybrid forms of expression, combining communist ideology with elements of Western (material) culture, past and present, feudal and capitalist. In addition to a myriad of consumer goods and a number of notable interior designs, this somewhat nostalgic and romantic 'medieval trend' materialised in a whole range of motion pictures. Many films in this cycle were musical productions – both short concert films for TV, featuring contemporary hits by local pop stars, often in jazz idiom, and theatrically released feature-length musicals, scored by equally young Estonian composers. This chapter explores a number of concert films, as well as two feature-length musicals – *Old Thomas Was Stolen* (1970) by Semyon Shkolnikov and *Don Juan in Tallinn* (1971) by Arvo Kruusement – that feature the Old Town as a spatial dominant and provide interesting insights into Soviet discourses of tourism and space, as well as popular music. Importantly, they formed a significant front of the Cold War where the West bombarded the Eastern bloc with products of capitalist entertainment, forcing Soviet authorities to allow increasing access to the images and sounds of this 'Eldorado of consumption' (Mattelart 1999: 113). Initially a strategy of social pacification that became crucial after the Prague Spring of 1968, in the long run, it contributed notably to the dissolution of the Soviet regime.

The aural and architectural landscapes of these films functioned as a two-way street. On one the hand, they served as display windows for exportation of Soviet achievements on the Baltic periphery to potential tourists, mainly outside, but also inside the Soviet Union. On the other hand, for local audiences they operated as illusory, yet probably somewhat comforting, tours to a virtual space behind the Iron Curtain, importing cultural references the access to which in 'real' life was restricted. In both respects, these entertainment-oriented audio-visual utterings are escapist and utopian in a sense described by Richard Dyer, 'offer[ing] the image of "something better" to escape into, or something we want deeply that our day-to-day lives don't provide' (Dyer 2002 [1977]: 20). Just as musical numbers in Hollywood productions, these sound- and cityscapes present 'ideal, joyous feelings, above all happiness' (Dyer 2012: 31). Importantly, according to Dyer, entertainment does not construct models of a utopian world; rather, it presents 'what utopia would feel like' and thus 'works at the level of sensibility' (Dyer 2002 [1977]: 20). However, this 'sensuous' utopia was not completely unproblematic, as it appears that this 'something better' at which they pointed was not the bright communist future prescribed by the Soviet authorities; rather, the proposed (escape) routes, as articulated in sound and images, directed, spatially, towards the (capitalist) West and, temporally, towards the feudal and bourgeois past, although in both cases without much critical socioeconomic consideration. In the process, these apparently rather trivial pieces of entertainment became involved in the construction of the sense of Estonian nationhood and facilitated the development of urban culture in a country and a nation that had been primarily rural and agricultural until the Second World War.

Tourism and the Old Town: towards the 'medieval trend'

From the late 1960s and early 1970s onward, as the hard-currency debt of the Soviet Union swelled, tourism became an indispensable source for obtaining convertible currency and as such essential for the functioning of the Soviet economic system. Alongside many Soviet cities, from Novgorod to Yerevan, from Kiev to Alma-Ata, from Riga to Tbilisi, Tallinn was included to the chain of attractions that the Soviet central tourism agency Intourist marketed to foreign tourists (see, e.g., Hall 1991: 37, 81). This practical, tourism-related cause was one of the key factors contributing to the massive popularisation of the medieval imagery of the Old Town. At the same time, the Old Town also came to signify a sense of national identity and resistance to the Soviet cultural discourse, distinguishing the local culture from the imposed Soviet regime and values, setting this unique historical environment in opposition to the official architectural paradigm of international modernism, which had neutralised local idiosyncrasy. Regardless of this, the Soviet system managed to 'colonise' not only Tallinn but countless other 'borrowed plums' from the history of Russia as well as of all other Soviet republics, by craftily weaving the material crust of the seemingly ideologically conflicting heritage into a single international cultural texture. This process, replacing the old meanings and functions with new ones and petrifying an everyday space into a frozen and consumerist collection of objects, was very much structured along the lines of the 'tourist gaze' (Urry 1990; see also Urry and Larsen 2011), and is especially evident in the cine-representations of the Old Town.

Estonian soundscapes: a Soviet enclave of Western-ness

Throughout the Soviet period, the Estonian landscape of popular music stood out as idiosyncratic, meaning more Western-oriented than in most other Soviet regions. This is testified by the trajectories of jazz, pop, and rock music, as briefly summarised below.

While the oft-repeated Soviet maxim, 'Today he is playing jazz and tomorrow he'll betray the Motherland', certainly holds some truth, the actual history of jazz in Soviet Union, and especially in Estonia, was more complicated, ranging from 'acceptance to full prohibition' (Reimann 2011: 28). However, the Estonian jazz scene flourished even during the darkest days of Stalinism when jazz was generally regarded 'as a product of bourgeois decadence and Jewish capital' (Atkins 2003: xxvi). During those postwar years 'the jazzmen of this tiny Baltic nation ... set the standards of creativity to which Russians were later to aspire in the post-Stalinist era' (Starr 1994: 230). Yet by the time most of the musical films under discussion here were produced, jazz was already, to a large extent, absorbed into the 'official' discourse of the Brezhnev era, as demonstrated by the fact that by 1966, the state record company Melodiya was releasing jazz records (Beumers 2005: 200).

In the context of pop music, it is important to emphasise that Estonia, and in particular its northern and western coastlines, were unique in the Soviet Union as they could be reached by an exceptional number of foreign radio transmissions. In addition to Voice of America, Radio Free Europe, and Radio Liberty, state-sponsored Western stations that specially targeted the audiences behind the Iron Curtain, and broadcasted popular tunes 'to ideologically destabilise the Soviet bloc regimes' (Mattelart 1999: 107), many North- and West-Estonian music fans tuned in to a number of official and pirate radio stations in Finland,[1] Sweden, and even as far as the UK (Salumets 1998: 61). In the mid-1960s, this led to the establishment of first pop bands in the entire Soviet Union. The Estonian groups, playing covers of popular Western hits and emulating their style in their original tracks, attracted more audiences than their Russian-speaking counterparts. Later, local progressive rock bands and in the 1980s also punk bands became famous throughout the Soviet Union with their original hit songs, making Estonia the most exciting music scene (though fairly isolated) in the USSR, as testified by Artemy Troitsky, a prominent Russian pop music expert (Troitsky 1987: 75). Among other things, this development was a result of Yuri Andropov's anti-rock policy, thanks to which the Western groups were replaced with the Soviet West, such as Estonians.

Troitsky's testimony of his first visit to Estonia is quite telling; he writes, 'my first visit to Estonia, at the end of 1975, was a real shock. ... everything there was different – they had rock traditions and a real rock "culture" ... The best thing about Estonian rock was the variety of styles ... It was all unbelievably impressive and unexpected – like Tallinn's Latin script and *Gothic architecture*. The dazed Russians in attendance walked around thoughtfully whispering to themselves about "the West ..." ' (Troitsky 1987: 38; my emphasis). Taking a cue from Troitsky's memories from the mid-1970s, emphasising the vivid first impressions of both music and Gothic architecture, the remaining part of this chapter traces how the music of Western origin and influence (although more jazz and pop than rock) intertwined with the retro-medieval spatial setting in the musical films of the 1960s and early 1970s.

Singing in the Old Town

The majority of short concert films, primarily made for TV, produced by Estonian Television Film, feature no coherent plot and are composed of popular tracks connected by a certain theme, recurring motif, or character. As such, they provide a parallel with the early proto-musicals or revues, 'programs of numbers with little or no narrative linkage between' (Bordwell and Thompson 2004: 123). Quite often, the Estonian productions follow the city-symphonesque temporal structure – from sunrise to sunset. Their approach to spatial representations draws principally on touristic pictorial syntax, shaped by the 'tourist gaze', by displaying a sequence of static, often sharp-angled shots of the 'gem objects' of the Old Town, drenched in sunshine and in a vacuum devoid of human figures (see also Näripea 2004).

The development of mise-en-scène of musical shorts between 1960 and 1970 demonstrates quite vividly the trajectory of the gradual popularisation of the 'medieval trend'. The earliest signs of the budding fashion can be detected in *Sound of the Day* (*Helisev päev*/*День полный песни*, 1960), a concert programme featuring the Estonian Radio Variety Orchestra and performing exclusively Estonian music, although marketed for local as well as Union-wide television audiences, as testified by its Russian titles, voiceover, and introductions to the individual tracks. The 40-minute-long programme anticipates the forthcoming wave of medievalesque retro in its title sequence and in the mise-en-scène of a single song. In the introductory sequence, the viewers are presented with brief establishing shots of early-morning Tallinn Old Town; its iconic silhouette as seen from the seafront; roofscapes and contours of its towers and spires immersed in the rays of rising sun; scenic perspectives of empty streets gradually filling up with people rushing to work; images of the Town Wall and its towers, most importantly the Long Hermann where the (Soviet) Estonian flag is hoisted daily, making it the ultimate symbol of power (and probably a sore reminder of the lack of independent statehood during the Soviet period). The opening song, *Hello, Day* (*Tere, päev*), illustrated with these images, is performed by a trio of female singers, including Heli Lääts, one of the ultimate stars of the period. The track was written by Uno Naissoo, a legendary Estonian composer and musician known for his love of jazz,[2] and lyrics by Heldur

Karmo, an equally prolific historian and theoretician of jazz, both of whom have contributed greatly to Estonian popular music as well as film soundtracks. The light-hearted optimism of the tunes and the lyrics matches perfectly the atmosphere of the images. The same can be said of the *Spring Story* (*Kevadine lugu*, music by Arne Oit, lyrics by Karmo), a song performed later in the programme by Georg Ots, undoubtedly the most revered singer of the period, mainly known as an opera soloist, who was immensely popular throughout the Soviet Union and also in neighbouring Finland. The set decoration for the *Spring Story* consists of a hand-drawn 'matte', a heavily stylised image of a medieval-looking street, complete with narrow façades of high-gabled buildings and a gas lantern attached to a wall, both of which were typical of Tallinn's Old Town. The gas lantern, in particular, became a shorthand for the 'medieval trend', being widely utilised in both interior and souvenir design, as well as in the graphics of Tallinnfilm's logo. Without openly challenging the Soviet ideology or approved modes of representation, the involvement of musicians (Naissoo, Oit, Karmo) who remained true to their sympathy to jazz at a period when it was disliked by Khrushchev's administration testifies of an implied desire to retain a certain sense of cultural difference, even if it does not manifest in the programme's predominantly rather conservative and symphonic musical idiom. That said, it is important to note that the programme was cautiously peppered with a few jazzy interludes, as demonstrated by the *Spring Story*.

In the mid-1960s, the USSR State Committee for Television and Radio Broadcasting commissioned a series of concert programmes from Estonian Television, known as the Melody-series, directed by Astrid Lepp between 1966 and 1969. This annual series, presenting the crème de la crème of Estonian pop music (then called *estrada* or 'variety music'), is significant in the development of Estonian musical film because it gradually introduced a plot-driven structure that replaced the quite arbitrary sequence of musical numbers that was typical before, including in *Sound of the Day*. Again, the songs, some of them still strikingly popular today, were typically written and performed in Estonian and by Estonian artists.[3] The series as a whole testifies to the steady popularisation of the 'medieval trend'. While, compared to *Melody-66*, the 1967 instalment presents more numbers staged and shot in the Old Town, only with *Melody 68* one can talk about a genuine breakthrough in terms of thematic consistency and dynamic treatment of space. In *Melody 68*, the filmmakers wander through the urban landscape quite freely and extensively, concentrating especially on the architectural environment and historical layers of the Old Town. Narratively, the individual tracks are tied into a more or less coherent whole by a main character, a man roaming the city in search of entertainment. First disappointed by the closed doors of various venues and signs announcing that the performance is already sold out, the man finally arrives at a tavern that provides a welcome relief to his pursuit of pleasure. The programme opens with a song that has become one of the most memorable of the entire era – *Miniskirt* (*Miniseelik*, 1967, written by Arne Oit, lyrics by Heldur Karmo), talking about a man who ignores the advances of a colleague until she starts wearing a mini dress – an iconic piece of clothing 'invented' by André Courrèges and Mary Quant that peaked in popularity in the 'Swinging London' of the 1960s and also reached behind the Iron Curtain. While the miniskirt was undoubtedly one of those Western commodity items that signified the 'Western-ness' of Estonia, the Old Town is equally important in this respect. In addition to the street scenes, functioning as a sort of a prelude, where the boldly jazzy tunes are framed by both contemporary architectural configurations and the narrow cobblestone streets of the Old Town hemmed with high-gabled façades, the film's main setting is the tavern where the pleasure-seeking protagonist finally finds his relief. Providing a background for several musical numbers, most of which are smoothly jazzy, with occasional schlager-like elements, this space, the vaulted structure of which creates a medievalised milieu, complemented with modern furnishings (a mixture then highly fashionable in Tallinn's entertainment venues), undergoes an unexpected metamorphosis, replacing the present with the past. The young couple whose quarrel had been the subject of the preceding song (*The End of a Love* [*Lõppenud on arm*], music by Ülo Vinter, performed by Helgi Sallo and Illart Orav), as well as other people who had earlier stared into their glasses, smoking, and looking bored, suddenly appear in historical costumes and the jaded hang-out turns into a spirited carousal. The first part of this episode is represented as a sequence of tableaux vivants, inspired by the pictorial world of the seventeenth-century Dutch Golden Age painting. In particular, one of the shots imitates Rembrandt's famous self-portrait with Saskia, *The Prodigal Son in the Tavern* (1635–1636). As their transfigured costumes also signal the seventeenth century, it is perhaps not entirely accurate to talk about 'medieval trend' here. Yet it is difficult to ignore the obvious associations with the nostalgic-retro currents of this trend in a more general sense, even more so as the inscriptions that have appeared on the walls of the tavern use Gothic script, which was one of the most emblematic features of this approach. A slightly comic number, *A Hint to the Drinker* (*Vihje viinavõtjale*, written by Valter Ojakäär), performed by the protagonist, now in historical attire, is followed by several melancholic, intimate ballads that narrate the process of reconciliation between the quarrelling couple. In the final sequence of the film, another metamorphosis, just as sudden, returns the protagonist to the present time, succeeded by a demonstration of the newest achievements of contemporary architecture, portraying these modern 'pearls' in trendy and dynamic low-angle cinematography. The final track, *I Am Fascinated by the City* (*Mind võlub linn*, written by Felix Mandre, lyrics by Heldur Karmo, performed by Illart Orav), is upbeat in mood, and its jazzy tunes find a visual parallel in the freely roaming camera that, in addition to contemporary architecture, captures the happily reunited lovers, strolling towards their (presumably) bright future. While the jazz-centred soundtrack testifies of the 'normalisation' of jazz as an approved form and the happy ending calls for reconciliation (with the Soviet present), the earlier, quite bizarre detour into history suggests a yearning for the cultural 'home' in the West.

Old Thomas Was Stolen

Surveys conducted among the cinema audiences in the 1970s demonstrated that the musical was one of the most popular genres, alongside melodrama, romance, comedy, crime, science fiction and the epic (Stites 1992: 169–170). Research also revealed that audiences were most attracted to contemporary topics, fast tempo, spectacularity, and simple 'invisible' editing patterns. Most of these attributes characterise all films under discussion here. Yet the fact that many productions made in Soviet Estonia in this period are drawn, in one way or another, to representations of history and historical spaces, seems to suggest a desire to depart from certain contemporary Soviet values. At the same time, the promotion of musicals, especially of the comedic variety, formed a firm part of Goskino's (the USSR State Committee of Cinematography) official line, with the intention to provide a counterbalance to more 'serious' genres, in order to pacify society, and, equally importantly, to increase income (Stites 1992: 171), including from abroad. For this end, the historical imagery of Tallinn's Old Town, which was easily and inexpensively available, offered an attractive frame of production, which, however, was often combined with contemporary themes and settings. *Old Thomas Was Stolen* (*Varastati Vana Toomas*, directed by Semyon Shkolnikov), perhaps Tallinnfilm's most notorious 'revue film', made in the 1970s, is a perfect example of this trend, integrating the old and the new. Commissioned by USSR Central Television 'for colour television' (which was then still an innovation) and 'for exchanging with foreign countries',[4] this musical comedy was quickly denounced as the biggest disgrace of the contemporary Soviet Estonian cinema.[5] As a commission targeted to foreign audiences, the film was ascribed 'a function of tourism promotion',[6] which in turn called for the 'demonstration of the beauty, honour and glory of Tallinn'.[7] Although USSR Central Television ultimately backed out of the project due to excessive cost, the film retained its commercial dominance of tourism marketing. Subsequently, it failed in both domestic and Union-wide distribution, and in 1973 was singled out as one of the worst films of the Soviet Union by the trade magazine *Soviet Screen* (*Sovetskii Ekran*) (Orav 2003: 522).

The protagonist of the film, Old Thomas, was a symbol of Tallinn, identified in the script proposal as the most popular 'person'.[8] However, as an emblem of the city, it was clearly of Soviet origin, soon acquiring the status of *de facto* 'master image' of Tallinn's tourism discourse. Essentially a weather vane put on top of the spire of Tallinn Town Hall in 1530, starting from the 1960s the figure of Old Thomas found its way to innumerable souvenir items, but also to culturally 'higher' representations, as demonstrated by his image on the cover of a catalogue for *Art in Estonia and Tallinn from Middle Ages to Today*, an exhibition which took place in Kiel at the end of the 1970s (Kuuskemaa 1978: 28). In addition, a street in Tallinn was named after him in 1963, as well as a restaurant and an orchestra (The Musicians of Old Thomas; Kallaste 1971); and he provided inspiration for several songs and films. This broad spectrum of popularity is commented on in *Old Thomas Was Stolen* by the fake Old Thomas who sings self-reflexively,

> By now Old Thomas has mastered new occupations
> Clad in knickerbockers and velvet
> I open games and shows
> And folk dance evenings
> As the symbol of the city
> A ferrous veteran
> I am the Old Thomas
> A beacon to the ships in search of land
> Directed by my moustache
> And this is art
> I am famous and honoured on the dry land
> Praised in a number of modern songs
> And by Tallinnfilm and by a desk lamp[9]
> – all of them of superior quality!

The 'authentic' Old Thomas, the protagonist of the film, is portrayed as a rather odd combination of an annoying watchdog of morality and a noble-minded gentleman. In the film, he typically serves as a vehicle of propaganda, pointing to the 'temporary' deficiencies of the communist society, such as littering, as well as ignorance and rudeness towards fellow citizens. More importantly, however, Old Thomas seems to function as a guardian of the Soviet people, protecting them from the dangerous Western influences, despite of the fact that the film itself is full of them.[10] For example, in a scene where an old lady is utterly shocked by the 'shrill and noisy music' (Kivilo 1971) of a jazz band, Old Thomas kindly offers her ear plugs, commenting that 'these helped me greatly when the town was bombed during the Livonian War' (1558–1583).

The sudden 'animation' of Old Thomas, the departure from his post at the Town Hall and his touring of Tallinn forms the structural axis of the film's somewhat indistinct plot. Chased by two 'archivists-numismatists' (Kivilo 1971) whose dialogue is largely made up of Latin sentences, his adventures loosely bind together a series of naively optimist pop song and dance numbers, all rather stereotypically Soviet in spirit, performed by politically correct characters (a janitor, a little girl, a female pop group, etc.). Due to the weakly tied narrative thread, the structure of the film, which again follows the familiar 'from sunrise to sunset' pattern, is closer to a concert programme than to a 'proper' musical, bringing to mind the later instalments of the *Melody* series (Figure 7.1).

Old Thomas Was Stolen opens at dawn, in the empty Town Hall Square – the symbolic centre of the Old Town (and the entire city of Tallinn) – where a young couple are romantically serenaded by Ivo Linna, one of the paramount pop stars of the day. The idyll, musically expressed as a light-hearted bossa nova, with lyrics that convey the freshness of spring – an almost grotesquely ironic reference in the wake of the era of stagnation – is abruptly broken when a gang of motorcyclists arrives. Their 'attack', staged to an instrumental jazz track as a variety dance number, rouses Old Thomas who comes to the rescue of the couple. Having scared them off by swinging his sword, the couple take Old Thomas on a tour of the city, launching a series of thematic performances. An act by a janitor, sweeping the cobblestoned streets, singing of his honourable and enjoyable profession in both Estonian and Russian, is followed by the performance of a jazz band on one of the observation decks of the Upper Town (Toompea), where the unmelodic screeches of the singers, as well as the adolescents bizarrely twisting to the (un)rhythms of these weird tunes, obviously serve as references to the undesirable influence of the capitalist West. Although

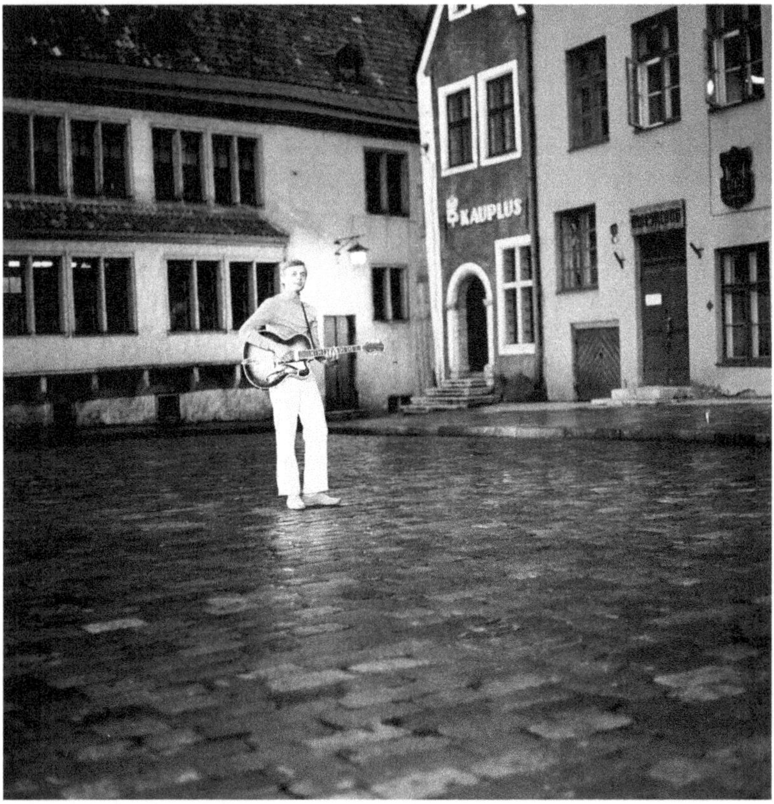

Figure 7.1 Ivo Linna in *Old Thomas Was Stolen*
Source: Photo courtesy of Estonian Film Institute and Film Archives of the National Archives of Estonia.

later an older man lectures Old Thomas about the value of 'true beat music', he remains unconvinced. The next number by a small blonde girl firmly re-establishes the Soviet ideal. The programme also includes a sea-themed track by the popular female group Laine (Wave) and a song and dance number about 'regular space transport' scheduled to travel between Earth, Venus, and Mercury, an unmistakable reference to the Soviet achievements and aspirations in space exploration and, by implication, colonial conquests. The culmination of the day arrives with the above-cited song by the 'fake' Old Thomas at the iconic Tallinn Song Festival Grounds, a place where every fourth year the Estonian national ethos was and still is fortified with the tradition of choral singing, the emblematic 'cradle' of Estonia as a nation, and where in the late 1980s the ground-breaking events of the Singing Revolution were played out as part of the process that led to the dissolution of the Soviet Union.

The mostly upbeat playlist is complemented by picturesque images of the Old Town, views that were firmly established in numerous guidebooks and coffee-table books, presenting the 'gem objects' of the Old Town in a fragmented, museumised, and sun-drenched mode of representation. Adding to these a few shots of contemporary architecture, the film comes across as an audio-visual tourist guide, under the headline 'a day in Tallinn, the capital of beautiful and progressive Baltic resort', with subtopics such as 'Tallinn – a City of Sea' and 'Tallinn – a City of Towers'. These phrases were repeated with monotonous consistency in almost every tourism-oriented publication of the era. The narrative covers not only historical treasures of art (special attention is paid to the extensive collection of coats of arms at the Dome Church, accompanied by organ music) and architecture (the churches, the Town Wall, several iconic streets and squares), but also modern achievements of construction (especially the newly erected Viru Hotel recently built in the centre of Tallinn by a Finnish construction firm to accommodate the increasing number of foreign tourists; and the vast expanses of 'Soviet bloc blocks' intended to provide housing to the masses of workers imported to this western periphery from other parts of the Soviet Union). In addition to these historical and modern 'pearls' the film includes a visit to a Song Festival, vividly demonstrating how the Old Town – as a heritage of feudal and capitalist nature – was integrated into the Soviet cultural paradigm in accordance with the socialist realist tenet 'national in form and socialist in content'. This relatively painless integration was primarily made possible by the general nature of rules and tasks prescribed by the tenets of socialist realism. For example, the fact that the 'national form' was defined as a 'living and developing phenomenon, shaped by history' (Taev 1945) made the concept flexible, allowing construing as national not only the pseudo-ethnographic imagery of folk art, but also the elements characteristic of Gothic architecture, such as pointed arches and finials. These were treated as the rural and urban aspects expressing the 'psychic idiom' (ibid.) of one and the same nation. The integration of the past and present and the figure of Old Thomas as a watchdog of morality, in turn, activate two main principles of the socialist realist canon: the didactic nature of the art and its mandatory association with the contemporary era (e.g. Hubel 1940: 854). Yet the film also contains numerous elements of contemporary Western youth culture, which testifies to the possibility of a multi-layered and sometimes contradictory conglomeration of paradigms even in the frame of reference that at the first glance comes across as quite exclusively Soviet in nature. This conflicting, even paradoxical fusion demonstrates that the entire 'medieval trend' was based on a complicated dialectic of adaptation and resistance. In this case, however, the elements of Soviet discourse predominated over those signalling the Western-ness, which is perhaps the reason for the film's ultimate failure.[11] It simply did not satisfy the audiences craving for 'genuine' experiences of the West, coming across as an impotent attempt to repackage the Western product.

Don Juan in Tallinn

Don Juan in Tallinn, a musical comedy directed in 1971 by Arvo Kruusement, was produced by Tallinnfilm and commissioned by the Soviet tourism authorities. While the film's authors have deemed it embarrassing hackwork, the successive generations of audiences continue to appreciate its fascinating camp qualities. In addition to a bizarre marriage of Gothic cityscape and contemporary fashion, the pop score written by young Olav Ehala, then a first-year student of composition and now a grand old man of Estonian pop music, contributes to a significant extent to its enduring popularity. Even though it was an officially approved product of the state-sponsored Soviet culture industry, aimed to attract foreign tourists and their hard currency to the Soviet Union, and also seen by more than 12 million Soviet viewers in the first year after its release, there is something in the film that suggests a certain deviation from the acclaimed values and ideals of Soviet society 'proper', thus revealing the essential hypocrisy of the system.

Don Juan in Tallinn is an adaptation of an earlier popular theatre play, *Once upon a Time in Sevilla* (1964), written by Samuil Alyoshin and stage directed by Kruusement. The play was obviously inspired by the stories of seduction related to Don Juan, the Seducer of Seville, a hero-villain of Spanish folk legend. In the legend, Don Juan is a famous lover and scoundrel, who has made more than a thousand sexual conquests. In Estonian rendition, however, Don Juan is a young lady disguised as a man, and instead of Seville, she arrives in Tallinn with her servant Florestino, creating excitement and confusion among the female population, as well as piercing the heart of the Commander who sees through her camouflage. The plot includes most of the key moments and characters of the original legend, interpreted in a parodist mode. The film stands out for its production design, visual excess, and hybridisation of styles which invite it to be read as a product of camp sensibility, characterised, according to Susan Sontag, by its 'love of the exaggerated' (Sontag 1999 [1964]: 56), which often tends to mean 'bad taste', and by a special emphasis on 'texture, sensuous surface, and style at the expense of content' (ibid.: 55). Indeed, the film places excessive emphasis on the surface of (audio)visuality and style at the expense of substance. Yet this anti-serious attitude and light-hearted-ness facilitates an appreciation of it as a kind of reflection of the numerous absurdities and discrepancies inherent to Soviet society at large. Most importantly, the entire mise-en-scène is heavily invested with signs of Western popular culture and capitalism, the arch-enemies of the socialist hemisphere, as demonstrated, for instance, by the highly fashionable (and thus commercialised) 'hippiesque' costumes and trendy interior designs. The apparent promotion of consumerist values acquires clearly ideological confrontational connotations within the framework of Soviet-Western opposition, surpassing, although under the disguise of irony, the officially acceptable limits of 'soft' Soviet consumerism. Its genre also situates it in the Western paradigm, because instead of the earlier Soviet-produced examples, the filmmakers decided to craft it after surprisingly recent American examples, namely *The Sound of Music* (1965, directed by Robert Wise), *Oliver!* (1968, directed by Carol Reed), *Hello, Dolly!* (1969, directed by Gene Kelly), and *Sweet Charity* (1969, directed by Bob Fosse), which they were able to see by a special request to the Committee of Cinematography of the ESSR. Precisely this consistent and intentional inclination towards the world on the other side of the Iron Curtain encourages one to interpret the film as a confrontational phenomenon in terms of national and cultural relations between Soviet Estonia and its Eastern conquerors.

The most forceful source of camp and the clearest manifestation of countercultural sensibility in *Don Juan Tallinn* is certainly the mode in which it represents history and environment. Again, this can be attributed primarily to the style, both visual and musical. In *Don Juan in Tallinn*, the subversive currents become especially notable in the somewhat strange and ambiguous treatment of historical (un)reality. On one hand, the (stylistic) aspects of the costumes and the design of the settings, as well as the 'medievalising' qualities of the locations appear to suggest certain historicity, but on the other hand, these clues are clearly stylised, mixed with contemporary, fashionable elements, resulting in a temporal vagueness. At times, the film seems to be particularly self-conscious about its intentionally off-beat construction of spatiotemporality. For instance, one of the first scenes features a 'medieval' interpretation of a beer vending machine and a street sign allowing the parking of horses at a hitching post; in the tavern, the waiter introduces a list of available beverages, most of them well-known Estonian brands of the 1970s; the simulated duel between Don Juan and the Commander is commented by Florestino's voice coming from a tape-recorder, while Florestino himself is sitting on a treetop, eating a loaf of bread and drinking milk from a bottle (just like the ones sold at grocery stores in the 1970s). The design of Don Ottavio's seaside 'palazzo' not only mixes ethnographic and modernist imagery, following the ultimate Soviet principle of 'national content, socialist form', but does it in a highly fashionable way, citing a particular contemporary trend of summer cottage design and thus relying on an aspect of *local* tradition and its modern-day interpretations as well as a predominantly Western frame of reference, which, once again, indicates the position of the Baltic states as the so-called Soviet West (cf. Kalm 2001: 369–374) (Figure 7.2).

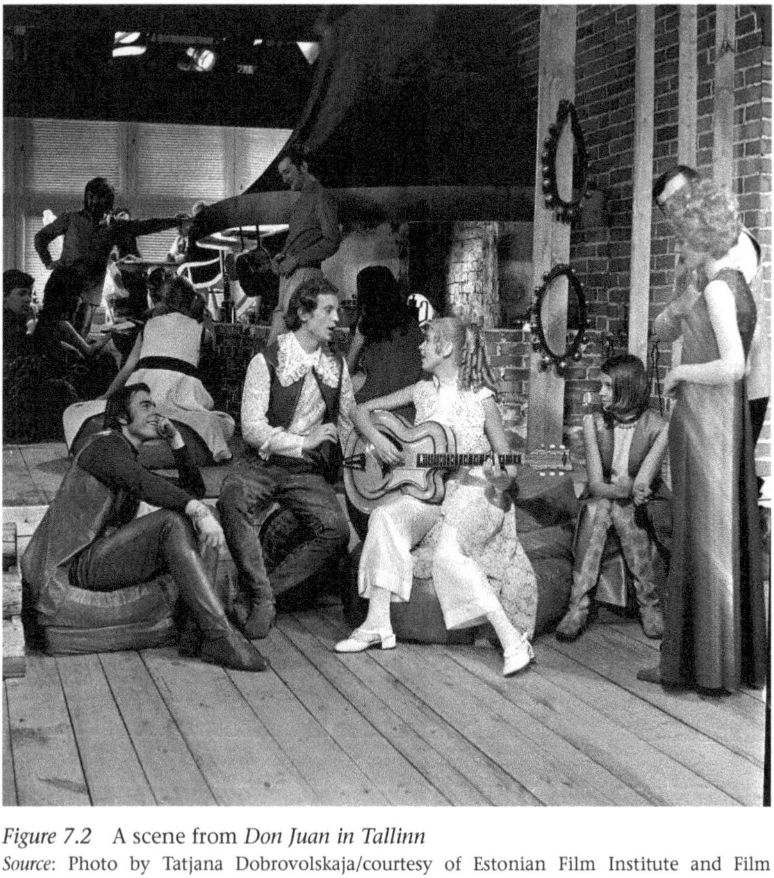

Figure 7.2 A scene from *Don Juan in Tallinn*
Source: Photo by Tatjana Dobrovolskaja/courtesy of Estonian Film Institute and Film Archives of the National Archives of Estonia.

A similar sense of Western-ness and hybridisation permeates the film's audioscapes. The original score was written by Olav Ehala, at the time only a 20-year-old first-year student of composition at the Tallinn State Conservatory. Previously, he played in a band called the Crystals (Kristallid) that preferred rhythm and blues to the Beatles, unlike the most notable underground pop (cover) groups of the 1960s (the Juniors/Juuniorid, the Optimists/Optimistid, the Contrasts/Kontrastid, the Aurora Borealis/Virmalised, etc.). Ehala's score for *Don Juan in Tallinn* is striking for its stylistic and spatiotemporal hybridity. On one hand, the sound is entirely contemporary, mixing pop threads with those of jazz that by the mid-1960s had been more or less accepted by the Soviet authorities. On the other hand, Ehala's music clearly betrays his love for Baroque music and Viennese classicism, which he himself has also admitted (Remme 2005: 7). His synthesis of various musical genres – pop, jazz, classical – reflects the cultural force fields characteristic to the Soviet Union: the underground pop, the tolerated jazz, and the lauded classical music, all Western in origin, testifying not only to Estonia's striving towards the cultural West, but also of the actual closeness of the two ideological hemispheres on both sides of the Iron Curtain.

The hybridist treatment of historical references, both visual and musical, of *Don Juan in Tallinn* indeed seems to undermine the Soviet establishment in certain respects, mostly through stressing the tone of a certain cultural difference, Western-ness, which challenges the boundaries of what was tolerated as 'measured doses of distraction and escapism in foreign lands' (Mattelart 1999: 111). Importantly, the film does so from *inside* the establishment, thus following Michel Foucault's idea that the platform of opposition to the power is in the network of power-relations themselves and not anywhere outside (see, e.g., Foucault 1995 [1975]: 26–27). As such, *Don Juan in Tallinn* reflects the ambivalent nature of the entire 'medieval trend' so prevalent in cultural fabric of the late 1960s and early 1970s in Soviet Estonia. The Middle Ages and its environmental equivalent, Tallinn's Old Town (where most of the location shooting of *Don Juan in Tallinn* was done), became the arena of cultural, ideological and economical negotiations, simultaneously displaying the representational ambitions of the Soviet system and the counter-directional aspirations of self-determination of the local, suppressed community.

Conclusion

The popular enunciations examined in this chapter bear witness to the process characteristic to the entire Soviet bloc – they served as swinging battlefields on the entertainment front ing the Cold War. Their success (or failure, in one case) demonstrated that popular music, which was intentionally broadcast by Soviet-oriented Western radio stations, was indeed an 'ideological weapon in the arsenal of democracy of the greatest effectiveness' (Cedric Larson, quoted in Mattelart 1999: 107). From the Soviet point of view, they became undesirable *loci* of resistance, as their instrumentalisation for anaesthetisation of social protest (Mattelart 1999: 111) was unsuccessful in the long-term, contributing to the destabilisation and ultimate disintegration of the regime. Hence, from the Western perspective they constituted a victory of an important battle in the propaganda war, as they signalled the adaptation of Western values (if in a naïve manner) behind the Iron Curtain, forcing Soviet authorities to ever-greater concessions in their principal ideological tenets. Finally, from the local, i.e. Estonian viewpoint, these *prima facie* light-hearted pop products functioned as sites where a sense of national belonging and local urban culture was produced. Pushing the boundaries of the limits set by Soviet authorities, even without seriously challenging them, these audio-visual utterances expressed signs of adaptation, not active rebellion, but still acted as kinds of *loci* of cultural difference, which ultimately proved to facilitate not an illusive, but an *actual* escape from the Soviet regime.

Notes

1. It is also important to note the immense influence of Finnish television on Estonian popular music and audio-visual culture.
2. Remarkably, he established the Swing Club, a jazz band, in 1948, at a time when jazz was prohibited in the Soviet Union; a year later he also launched the first (Soviet) Estonian jazz festival in Tallinn (see, e.g., Starr 1994: 231–232). Later, teaching musical theory at the Tallinn State Conservatory, from which he had graduated in 1952 and where he worked throughout his professional life, he became the founder of the department of pop-jazz (under the politically correct title of *estrada*) in 1977.
3. One of the notable exceptions being the Estonian cover of *Yesterday* (written by Paul McCartney, 1965), performed in *Melody-66* under the Estonian title of *Eile veel* (lyrics translated by Heldur Karmo in 1966) by Tiiu Varik only a year after its original release. On 'Beatlemania' in the Soviet Union, see, e.g., Mattelart 1999: 108–109).
4. Documentation of *Old Thomas Was Stolen* (1968–1971), The National Archives of Estonia, ERA.R-1707.1.1127, p. 8.
5. Kaljo Kiisk, the head of the Soviet Estonian Union of Filmmakers, defined it in his presentation at the third congress of the Union in 1971 as a 'completely empty phenomenon ... entirely outside of the limits of art' (a shorthand report of the congress; ERA.R-2245.1.124, pp. 49, 55).

6. ERA.R-1707.1.1127, p. 41.
7. ERA.R-1707.1.1127, p. 9.
8. ERA.R-1707.1.1127, p. 5.
9. This is a reference to a night lamp in the shape of a gas lantern, designed by Bruno Vesterberg in 1966, which became (and still is) a genuine cult object.
10. Accusations of 'excessive Western-ness' were also voiced by some members of the studio's Artistic Council in their assessments of the film (ERA. R-1707.1.1127, p. 9).
11. During its first year, the film attracted 6.4 million viewers, while *Don Juan in Tallinn* scored 12.2 million admissions Union-wide (Vabamägi and Teinemaa 1989: 66).

Works cited

Atkins, E. Taylor (2003). 'Towards a global history of Jazz', *Jazz Planet*. Ed. E. Taylor Atkins (Jackson, MS: University Press of Mississippi), pp. xii–xvii.

Beumers, Birgit (2005). *Pop Culture Russia! Media, Arts, and Lifestyle* (Santa Barbara, CA: ABC-CLIO).

Bordwell, David and Kristin Thompson (2004). *Film Art: An Introduction*, 7th ed. (New York: McGraw-Hill).

Dyer, Richard (2002) [1977]. 'Entertainment and Utopia', in *Hollywood Musicals: The Film Reader*. Ed. Steve Cohan (London: Routledge), pp. 19–31.

Dyer, Richard (2012). *In the Space of a Song: The Uses of Song in Film* (Milton Park and New York: Routledge).

Foucault, Michel (1995) [1975]. *Discipline and Punish: The Birth of the Prison* (New York: Vintage Books).

Hall, Derek R. (1991). *Tourism and Economic Development in Eastern Europe and the Soviet Union* (London: Belhaven Press; New York and Toronto: Halsted Press).

Hubel, E. (1940). 'Jooni sotsialistlikust realismist', *Looming* (8), pp. 854–865.

Kallaste, V. (1971). 'Miks Vana Toomas lööb hädakella?', *Noorte Hääl*, 20 February.

Kalm, Mart (2001). *Eesti 20. sajandi arhitektuur / Estonian 20th Century Architecture* (Tallinn: Prisma Prindi Kirjastus).

Kivilo, Vidrik (1971). 'Mille nimel "Varastati Vana Toomas"'?, *Sirp ja Vasar*, 26 February.

Kuuskemaa, Jüri (1978). 'Vana Toomas restaureerimislaual', *Kultuur ja Elu* (11), pp. 25–28.

Mattelart, Tristan (1999). 'Transboundary flows of western entertainment across the iron curtain', *Journal of International Communication* 6(2), pp. 106–121.

Näripea, Eva (2004). 'Medieval socialist realism: Representations of Tallinn old town in Soviet Estonian feature films, 1969–1972', *Koht ja paik / Place and Location. Studies in Environmental Aesthetics and Semiotics* 4. Ed. Virve Sarapik, Kadri Tüür and Eva Näripea (Tartu: Eesti Kirjandusmuuseum), pp. 121–144.

Orav, Õie (2003). *Tallinnfilm I. Mängufilmid 1947–1976* (Tallinn: Eesti Entsüklopeediakirjastus).

Reimann, Heli (2011). 'The ideological context for the study of Soviet Estonian Jazz', *Jazz Chameleon: The Refereed Proceedings of the 9th Nordic Jazz Conference August 19–20, 2010, Helsinki, Finland*. Ed. Janne Mäkelä (Helsinki: The Finnish

Jazz & Pop Archive; Turku: International Institute for Popular Culture), pp. 23–35, http://iipc.utu.fi/jazzchameleon/ (accessed 16 September 2014).

Remme, Anneli (2005). 'Vastab Olav Ehala', *Teater. Muusika. Kino* (1), pp. 5–13.

Salumets, Vello (1998). *Rockrapsoodia* (Tallinn: Eesti Entsüklopeediakirjastus).

Sontag, Susan (1999) [1964]. 'Notes on "Camp"', in *Camp: Queer Aesthetics and the Performing Subject: A Reader*. Ed. Fabio Cleto (Ann Arbor: University of Michigan Press), pp. 53–65.

Starr, S. Frederick (1994). *Red & Hot: The Fate of Jazz in the Soviet Union*, 2nd edition (New York: Limelight Edition).

Stites, Richard (1992). *Russian Popular Culture: Entertainment and Society since 1900* (Cambridge: Cambridge University Press).

Taev, Karl (1945). 'Sotsialistliku realismi rahvalikkus', *Sirp ja Vasar*, 2 June.

Troitsky, Artemy (1987). *Back in the USSR: The True Story of Rock in Russia* (London: Omnibus Press).

Urry, John (1990). *The Tourist Gaze: Leisure and Travel in Contemporary Societies* (London: Sage Publications).

Urry, John and Larsen, Jonas (2011). *The Tourist Gaze 3.0* (Los Angeles: Sage).

Vabamägi, Edvi and Sulev Teinemaa (1989). 'Tehtud ja tegemata teod. Arvo Kruusemendi filmidest', *Teater. Muusika. Kino* (8), pp. 59–71.

8
Tourism and Heterotopia in Falco's Songs

Ewa Mazierska

The purpose of this chapter is to analyse the motif of tourism, as a form of exploring foreign places, in the work of Austrian singer and song writer Falco. In order to do so, I will refer to two concepts: 'tourism' and 'heterotopia', both used in the discourse on popular music. 'Music and tourism' covers such phenomena as promoting specific locations through the lyrics and music of the songs and pilgrimages to sites related to famous musicians such as Memphis, the site of memory of Elvis Presley or music festivals (Gibson and Connell 2005). However, not much is written about musicians engaging with the discourse on tourism, adopting the role of critical travellers which, I will argue in due course, is the case with Falco.

The concept of heterotopia appears most often in studies of specific music phenomena which are marked by hybridity. Josh Kun argues that American popular music is heterotopian because it reflects an influence of different music traditions, as opposed to being homogenous, what Kun describes as 'symphonic'. Kun welcomes this hybridity, arguing that it is a means to create a social utopia (Kun 2005: 1–28). Philip Bohlman compares two models of Jewish music culture: one pertaining to a shtetl, the other to the ghetto, where the former can be seen as homogenous, 'symphony-like', the other heterotopian, cacophonic. Although the term 'ghetto' normally awakens negative associations, Bohlman sees the music of a ghetto in positive terms, as fostering social tolerance and development of art (Bohlman 2002).

We can talk not only about tourism as a specific activity, but also about the tourist gaze as a specific way of looking at reality. By the same token, we can talk about the heterotopian gaze. In the remaining part of this chapter I will attempt to establish whether Falco's songs and videos offer us an experience pertaining to that of the tourist or heterotopian gaze. I will do it in two parts, first by identifying common features of his songs, second by charting changes in his attitude to tourism throughout his first three records, made in the years 1981–1985 and their accompanying videos. I chose these three records, because they mark a distinct trajectory, from presenting oneself as a local man and artist to being a 'critical traveller'. I shall mention that by 'Falco', 'Falco songs' and 'Falco videos' I do not mean a private individual and a song created solely by Falco, but an effect of the text, on which Falco collaborated with music producers, videomakers, and many other artists.

Falco as a local and transnational artist

Falco (1957–1998) was the greatest Austrian postwar pop star, selling over 60 million records. His period of greatest success was in the 1980s, when his songs, such as *Der Kommissar, Rock Me Amadeus, Vienna Calling* and *Jeanny,* topped the American and European charts. His international success was the result of a combination of factors, such as intensification of the velocity of rock music and rock culture in the 1980s (Straw 1993), which opened up a space for incomers and outsiders, even those whose principal language was not English, as was his case. Another factor was the postmodern character of his music, which was the dominant idiom in the 1980s. It is difficult to find an artist who in such a relatively short career explored so many musical styles: funk, punk, disco, techno, rock, reggae, tango, bossa nova, Latino and Indian sounds. Not only did he navigate between different musical traditions as he matured as a musician, but mingled many musical languages in one piece. His song might begin like a waltz, change into a dynamic rock piece and finish in a jazzy way or yodelling. Falco was also both a singer and a rapper, and in a somewhat unusual way for a rapper, sang and rapped in one song. He was a rapper also in the sense of using samples from other works as well as city sounds, such as radio announcements or information passed by a flight attendant to the passengers on a plane. His songs, as other rap songs, can thus be described as 'sonic bricolages' (Potter 1995: 33); they inspired a more visual and spatial kind of listening than other types of songs and asked for videos to accompany them. This, indeed, happened and many of Falco's videos were very successful, for example those for *Jeanny, Junge Roemer,* and *Rock Me Amadeus,* which I will discuss in due course.

One should also mention the transnational character of the production of his music and his artistic identity: the producers of his greatest hits, the Bolland brothers, were Dutch; his manager, Horst Bork, was a talented German able to secure him contracts with leading music corporations and international tours, including in Japan (Mazierska 2014). Moreover, Falco was not the real name of my protagonist, but his artistic pseudonym; his real name was Johann (Hans) Hölzel. In due course, Falco also hid his 'Falco' persona behind another name, TMA, and in many of his videos played people who change identities, as in his most famous song and video, *Rock Me Amadeus*, which I will discuss in detail in the subsequent part of this chapter. All of his adopted names sound distinctly non-German, but rather transnational; 'Falco' bears similarity to 'falcon', but is not quite 'falcon'. The practice of using alter egos is common in popular music and especially in rap. One can think, for example, about the many names adopted by the most famous white rapper, Eminem. We can see this practice as a kind of 'spatialising self', which underscores the character of modern life, marked by mobility and urbanity, especially using new communications and media technologies, which allows one to be practically in many different places at the same time (Crang and Thrift 2000: 8–9).

In the late 1980s–1990s Falco's international career declined, for the same reasons which ensured his popularity before: the high velocity of music, a less attractive appearance, his failure to secure attractive deals, as well as burn-out and his inability to find a suitable successor to the Bollands, whose later productions did not match the hit-value of his earlier ones. Paradoxically, his career was revived by his tragic death in a car crash.[1] This ensured his legend as the one who 'died too young', along with many other pop stars and led to the speedy release of his last record, which proved very popular in the German-speaking world.

Falco can be construed in contrasting ways: on the one hand, as an artist strongly linked to his home town, Vienna, because he immortalised it in many of his songs, such as *Ganz Wien*, *Vienna Calling*, *Rock Me Amadeus*, and *Wiener Blut*. These are also the songs which brought him greatest popularity in his own country and globally. Even now he is remembered as a kind of ambassador of Vienna.[2] On the other hand, his songs have a uniquely cosmopolitan aura. A reader of Falco's lyrics is surprised by the multitude of locations listed in them: Zurich, Constantinople, Kenya, Hawaii, Panama, Himalaya, Jamaica, Arizona, Toronto, New York, Amman, to name just a selection. Some of these places reflect journeys undertaken by Falco , for example to the United States or Amman; others, like Panama or Jamaica, are symbols of specific lifestyles, marked by lying on a beach or taking drugs. The overall impression from this cacophony of names, many appearing in a single song, is that their protagonist is an affluent tourist who has no problem in going anywhere he wants. At the same time, mentioning so many places in the span of a four-minute song, as well as the way he describes some of these places, renders the journey between them unreal. This unreality was augmented by the versatile style of music used in an individual song, as already mentioned.

For these reasons Falco's songs bring to mind the idea of heterotopia as a space outside normal experience. Michel Foucault, who introduced this concept in 1984, later appropriated by postmodern thinkers, such as David Harvey (1990: 48–49), Kevin Hetherington (1997), and Daniel Jewesbury (2003)[3] defines heterotopias as 'counter-sites, a kind of effectively enacted utopia in which the real sites, all the real sites that can be found within the culture, are simultaneously represented, contested, and inverted. Places of this kind are outside of all places, even though it may be possible to indicate their location in reality' (Foucault 1998: 231). As examples of heterotopia Foucault mentions prisons and psychiatric hospitals, as well as colonies, brothels, and ships and lists five principles of heterotopias, of which the most relevant to my analysis is the fourth, namely heterotopias as heterochronies – they are linked to slices of times. A person who finds him/herself in a heterotopia is separated from the normal flow of time. For this reason, a cemetery appears to be a perfect heterotopia, because it begins with the loss of life (on cemeteries as heterotopias see Toussaint and Decrop 2013). Heterotopias also include theatre stages, film screens, museums, and libraries, because in them time is condensed by bringing together objects belonging to different spaces and times and different ontological orders. Heterotopias thus have much in common with touristy spaces, because in both of them time is bracketed off, as conveyed in the quote from Debord: 'The same modernisation that has deprived travel of its temporal aspect has likewise *deprived* [my emphasis] it of the reality of space' (Debord 1994: 120). But there is a difference between 'tourist gaze' and 'heterotopian gaze'. Those who direct and control tourist gaze and the tourist gazers themselves attempt to regard what they experience as 'authentic' and 'typical' for a specific time and culture, such as a typical French castle and – to move to an Austrian context – typical Austrian opera or Wiener schnitzel. The tourist gaze strives towards fixing and reification. The practices of tourism decontextualise, simplify, homogenise, romanticise, essentialise, and mystify the complex character of a specific historical reality (Albers and James 1988; Urry and Larsen 2011; Wang 2000: 160–171). 'Heterotopian eye', by contrast, sees the world in terms of rupture and the impossibility of imposing an order on the received world. This aspect of heterotopia is elaborated by Foucault in *The Order of Things*, where he maintains that

> *Heterotopias* are disturbing, probably because they secretly undermine language, because they make it impossible to name this *and* that, because they shatter or tangle common names, because they destroy 'syntax' in advance, and not only the syntax with which we construct sentences but also that less apparent syntax which causes words and things (next to and also opposite one another) to 'hold together'. This is why utopias permit fables and discourse: they run with the very grain of language and are part of the fundamental dimension of the *fabula*; heterotopias (such as those to be found so often in Borges) desiccate speech, stop words in their tracks, contest the very possibility of grammar at its source; they dissolve our myths and sterilize the lyricism of our sentences. (Foucault 2002: xix)

Falco's affinity in creating impossible spaces was recognised by some authors researching his lyrics. An Austrian scholar, Peter Ernst, wrote: 'In Falco's texts we encounter not real but possible spaces' (Ernst 2010: 126). Ernst bases this opinion on Falco's predilection to utterances in a conditional mode. Indeed, in the worlds conjured up in Falco's song events are typically waiting to happen. For example the ideal woman is not yet born, although she exists in the protagonist's imagination. On other occasions they appear to take place simultaneously in the past and the present, when the character rambles through the history of popular music, as in his *The Sound of Musik*. Alice Bolterauer analyses Falco's lyrics in the context of Jean Baudrillard's concept of simulacra as a space where the difference between reality and unreality gets blurred. Bolterauer argues that the subject of Falco's songs is never reality as such, but its representation, for example images conveyed by media clichés (Bolterauer 2010: 87–90); hence their unreality. Moreover, Falco had a predilection to places which for Foucault epitomised heterotopias due to being cut out from the normal flow of experience. He sung about ships (*Titanic*), brothels (*Munich Girls, Nachtflug, Königin of Eschnapur*), computer screens (*Data de Groove*) and cemeteries (*Monarchy Now*), on many occasions playing up their heterotopian aspects, especially the fact that each of the sites is a crossroads of various cultures and a place where people can play with their identities.

Furthermore, Falco mixed languages, most importantly German with English and Viennese slang with standard German, and added a sample of other languages, such as French, Italian, and Spanish. Such a mixture, in his own admission, fulfilled a realistic function: it reflected the way contemporary Austrians talked: the impurity of their German. However, while it is true that in the language of an average Austrian or a German-speaking person we find a fair dose of English, Falco's language play is more complex and idiosyncratic than a realistic representation requires or even undermines the songs' realism (Ernst 2010). Falco also indulged in the creation of neologisms. His strange words, especially accumulated on his *Data de Groove* record, such as 'expobeat' or 'anacond'amour' can be seen as miniature heterotopias, according to Foucault's claim that heterotopias 'secretly undermine language, make it impossible to name this *and* that, shatter or tangle common names, destroy "syntax" in advance, and not only the syntax with which we construct sentences but also that less apparent syntax which causes words and things (next to and also opposite one another) to 'hold together' (Foucault 2002: xix). I find these words capturing very well 'Falco's world' as through textual and musical means and their combination it conveys a struggle to keep a world which does not want to hold together, which proliferates and fragments.

The cacophony of languages and proliferation of neologisms points to the incompatibility of linguistic communities and the impossibility of bringing different cultures to the common denominator, which is also an idea conveyed by the content of Falco's songs. Such a conclusion can be seen as a reflection of Falco's experience as a real and cultural traveller. Although he travelled extensively for pleasure and work, often to exotic countries, such as Thailand and Japan, as well to the United States, and eventually settled in the Dominican Republic, his biographers claim that abroad he felt out of place and missed his native Vienna. But equally from a certain moment in his life he was not happy in Vienna; he was thus somewhat suspended and perhaps lost between locations and cultures.

Discourse on tourism from *Einzelhaft* to *Falco 3*

The first song in which Falco engages openly with tourism is *Hinter uns die Sintflut* from his first album, *Einzelhaft* (1982), produced in Vienna by Austrian music producer, Robert Ponger. It is regarded by many Falco fans as Falco's most 'authentic' record, in part due to being locally produced and containing songs with more of Falco's input than his subsequent songs. *Hinter uns die Sintflut* is written in the first person, but its content, as well as the style of music and manner of performance, which is different from the other songs on *Einzelhaft*, conveys a distance between the performer and his character. The person described in the lyrics, most likely an affluent, middle-aged German touring holiday resorts, reveals all the cardinal sins of a mass tourist and some additional ones. He is vulgar, because he goes to places not for the cultural pleasures they offer but because the sun is shining and he does not enter into any deeper contact with his destinations. Hawaii, Kenya, or Panama are interchangeable for him – he sees no difference between these countries.

Although tourists are encouraged to travel to see new things, these new things are in a sense the same, bringing to mind Guy Debord's comment that 'tourism is a by-product of the circulation of commodities; basically, tourism is a chance to go and see what has been made trite. The economic management of travel to different places suffices in itself to ensure those places' interchangeability' (Debord 1994: 120). Moreover, the protagonist of *Hinter uns die Sintflut* is conspicuous about his wealth and uses it as a pass to enjoy himself, indulging in the pleasures of the flesh. He shows colonial disrespect to his hosts, as demonstrated by the very title of the song which translates 'After us, the deluge' and lines such as 'Ich flog ohne Pyjama nach Panama' (I fly to Panama without my pyjamas). There is also a hint that the wealth of the protagonist was earned dishonestly and he might engage in tourism as a form of tax evasion. His description brings to mind the Reverend Francis Kilvert's, who as early as in the second half of the nineteenth century pronounced that 'If there is one thing more hateful than another it is being told what to admire and having objects pointed out to one with a stick. Of all noxious animals too the most noxious is a tourist' (Kilvert 1971: 79) and more recently Rob Shields, who argues that tourist experience is denigrated because it is marked by excess, resulting from an ostentatious breaking with the established routines of everyday life and being disruptive for the natives (Shields 1991: 73–116).

Music and the manner of performance of *Hinter uns die Sintflut* augment the impression of a shallowness and vulgarity of the travelling millionaire. It merges Calypso with a parodied style of prewar German singers, most importantly Hans Albers[4] and a chorus which is in a different style, more like other songs included on this record, as if to indicate that the true voice of the singer is different from that of the lampooned millionaire. The remaining songs from the record and, especially, *Auf der Flucht*, *Der Kommissar*, and *Zu viel Hitze*, shed additional light on *Hinter uns die Sintflut*. All of these songs are about moving, but their protagonists are people who run away from the police in their own city, and are increasingly confined, encircled, either because they are persecuted for drug abuse or because they engaged in illegal political activities, demonstrations, and riots. The luxurious and irresponsible behaviour of the millionaire tourist appears to be the obverse of some failed hippie revolution. According to the rock tradition on this record Falco himself comes across as a spokesman for the angry and economically and politically underprivileged youth.

A different approach to tourism is proposed on Falco's subsequent record, *Junge Roemer* (1984), also produced by Ponger, which, on Falco's own account, expressed his desire to flirt with the establishment, and to appeal to audiences beyond Austria and the German-speaking world. Videos were produced for all the songs included on this record, showing Falco in different locations, such as the United States, Vienna, and Egypt and in different periods. Most of the videos present a young and affluent man who is superficially at ease with himself, but the reality is often different – he is a fraud detached from the place in which he finds himself. From the perspective of tourism and heterotopia, especially interesting is *Tut-Ench-Amon* (*Tutankhamen*). The protagonist of this song is a tourist, but with a different attitude than the one in *Hinter uns die Sintflut*. While the former brings all places he visits to the lowest common denominator, perceiving them merely as vehicles to achieve carnal pleasures, the latter sees them in their difference and incompatibility and, consequently, has a sense of the absurdity of his position. He can be described as a 'heterotopian tourist', namely a traveller with an acute sense of being out of place, outside the normal flow of time. This is conveyed by phrases such as: 'feeling like a tourist', 'on a mission without a purpose, I wish I'd never been born'. Another sign of him being a 'heterotopian tourist' is testified by a saturation of the song with references to German colonial history. The lyrics begin with the words:

> Been and seen down flight from Chicago
> All aboard the African Queen
> I am on a night boat down to Cairo
> Put you in the picture – 1884
> I'm up to date on the literature
> But I have never been here before

The term 'African Queen', and the year 1884 point to the history of German colonisation in Africa. Eighteen-eighty-four is the date of the Berlin Conference when Germany claimed several colonies in Africa and *African Queen* is a title of John Houston's film of 1951, set during the First World War, which was, to a large extent, a consequence of

European colonialism. The title of the song, which is Falco's way of saying 'Tutankhamen', also points to the history of German discovery and appropriation of Egyptian treasures, which until now fill German national museums. By evoking this history, the protagonist of the song suggests that a German or maybe even any Westerner cannot or should not be a careless tourist. He has to look at Africa as a heterotopia, a site made of many histories which cannot be put together for unproblematic consumption because at each corner they reveal a legacy of conflict and violence. The heterotopian character of the space is also underlined by Falco looking at the visited sites as if they belonged both to the past and the present or even are suspended between different time frames. For example, the line 'If he looked any older, he might as well be dead', referring to an Egyptian man met during the trip, might allude to the fact that he sees in contemporary inhabitants of Egypt incarnations of the Egyptians from the times of the pharaohs. The song also criticises the natives who prey on tourists, and the tourism industry as based on a double exploitation of hosts and foreigners:

> There is a black boy on the corner trailing me behind
> Stalking like a vulture money on his mind

Tut-Ench-Amon thus deconstructs touristy space as a site which is in fact heterotopian, because it is fragmented, post(colonial), and impossible to enjoy in an unproblematic way. This point is reinforced by the title of the song, which fragments the name of the most famous pharaoh, changing 'Tutankhamen' into 'Tut-Ench-Amon'. Such change brings to mind Foucault's claim that 'heterotopias undermine language and shatter or tangle common names'. Indeed, the song produces on the listener (at least this listener) an effect of unease; Tutankhamen' fragmented into 'Tut-Ench-Amon' makes me question the reality behind the museumised culture of mummies and hieroglyphs and my largely superficial and unreflective attitude to the treasures of the past.

The video for the song reinforces these ideas. In a lavish car, Falco moves from the Schönbrunn Palace in Vienna to a landscape with pyramids, in a short cut which, of course, could not happen in reality. Then we see him inside a tomb, clad in a dark cape and dark make-up, as if he was a de-mummified 'tourist' from a distant epoch. The camera pans through what looks like the surface of tombs, covered in Egyptian hieroglyphs. Yet, among them we find pictures of cars – a sign that the authors of this work make no attempt to create a smooth touristy space but a parody of such a space by revealing fissures on its surface. And yet, in comparison with *Hinter uns die Sintflut* and other songs in *Einzelhaft*, which transmit an overt antipathy towards the tourists and take the side of the natives, *Tut-Ench-Amon* adopts the perspective of a tourist who, although self-ironic, regards his position as natural and impossible to escape. This is also a position attributed to the characters in other songs from this record, most importantly the title song, *Junge Roemer*. 'Junge Roemer', young affluent inhabitants of large towns are represented by Falco as alienated from their environment, feeding themselves on the past and engaging in excessive behaviour, as if they were tourists in their own town.

Falco's second record proved successful in Austria and, to some extent, Germany, but barely 'travelled' beyond the boundaries of the German-speaking world. As one reason for that, his manager lists the textual sophistication and ambiguity of the songs' lyrics (Bork 2009: 61). Consequently, his next record, *Falco 3* (1985) which proved the most successful in his career, was produced by Dutch producers, Rob and Ferdi Bolland, who were brought in to widen Falco's appeal beyond the borders of the German-speaking world. This meant, among other things, 'selling' him not only as a traveller who gazes at other cultures and people with his distinct postmodern, ironic eye, but also as an object to be consumed by a tourist gaze, somebody who is himself a tourist attraction. In the two most successful songs from this record, *Rock Me Amadeus* and *Vienna Calling*, these various positions of Falco, which are normally kept separate, that of a tourist and a native, are merged. As the very titles of the songs suggest, in them Falco is linked to his country, Austria or more specifically, to Austria at its most touristy, due to Wolfgang Amadeus Mozart and Vienna being Austria's best known treasure.

Given Falco's earlier critical engagement with tourism, the singer was not comfortable with the way he was linked to them. His biographers emphasise his initial opposition to *Rock Me Amadeus* (Bork 2009: 87–88, 100–103), precisely because the song conjured up Amadeus, and by extension, Falco, as a tourist attraction, through simplifying, homogenising, decontextualising and beautifying an object offered to a tourist gaze, in the way described by Ning Wang (Wang 2000: 160–171) and John Urry and Jonas Larsen (Urry and Larsen 2011). However, ultimately he gave in, convinced that the song would work and having no alternative strategy to achieve a commercial success. The Bollands not only wrote music for this song, but came up with an idea to capitalise on the immense popularity of Miloš Forman's film, *Amadeus* (1984) by playing up the fact of Falco and Mozart sharing (broadly speaking) the same nationality. The song performs a double decontextualisation and simplification. It takes Mozart out of his original, historical context and out of his cinematic context, translating him for audiences who might be familiar with both of these contexts but only superficially, and puts him in a new context, by using a different medium and performer.

Due to its title, which includes a well-known historical figure, *Rock Me Amadeus* invites comparison with *Tut-Ench-Amon*. However, these two figures are evoked differently in the respective works. The name *Tut-Ench-Amon* is in a heterotopian fashion fragmented. It poses the question whether we can recreate a historical figure in contemporary times. The whole song, as I have already suggested, is about the bewilderment of being a tourist, confronted with the past, which is commodified and mediated by the memory of colonialism. *Rock Me Amadeus*, by contrast, does not question who Amadeus was or how we can access him. Falco sings about Amadeus having debts, being a drunkard, and a womaniser. It also locates Mozart firmly in a specific place and time: Vienna and the year 1780. And yet, albeit more subtly, *Rock Me Amadeus*, like *Tut-Ench-Amon* proposes the idea that we can access the past only through the matrix of the present, using contemporary categories. This point is transmitted by such lines as 'war ein Rockidol' and 'no plastic money anymore'. However, such touristy decontextualisation does not cause bewilderment on the part of the transmitter of the messages, but appears to be taken for granted by him. I suggest that one reason why this song caused so much opposition during its time and subsequently, being regarded as kitschy and unsophisticated, is due to taking tourism and by extension historical discourse for what it is, namely for imagining the past as if it was a version of the present, in this case Mozart as a rock star. Historical books and films always do it. There is no other way to bridge the gap between the past and the present than by translating our experience into past experience; the alternative is to leave the past untranslated, fragmented, bewildering, and heterotopian. However, the authors of historical books and films are unwilling to admit that their historical works are in fact contemporary works. Moreover, a historical book or film has more time at its disposal to inculcate in the reader/viewer an idea that somebody from the past behaved in a contemporary way; time to provide the appropriate context. This is not the case with *Rock Me Amadeus*, which is only slightly longer than a standard song.

The idea that Mozart's past was like Falco's present is reinforced in the video version of the song, in which we get a montage of two scenes. One presents some members of eighteenth-century Viennese high society,

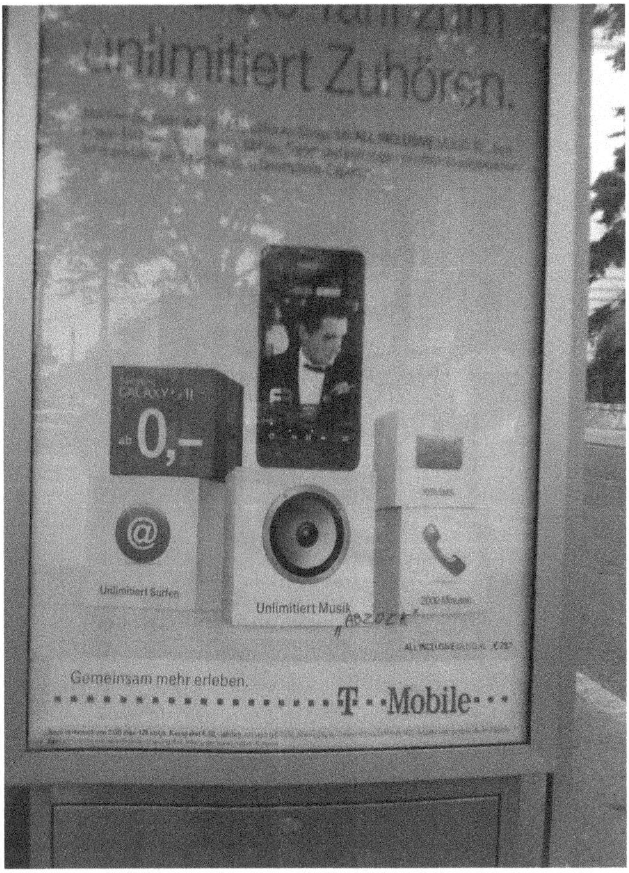

Figure 8.1 Falco's image used in the advert of T Mobile

dressed in formal clothes. Falco occupies a privileged position in this world. He alights from a coach, dressed in a tuxedo and a bow tie, in a way which marks him as belonging to contemporary times. The second scene shows Falco in clothes from Mozart's time and a colourful wig, but surrounded by contemporary rockers on motorcycles. Neither of the scenes thus belongs univocally either to the past or to the present and eventually the two scenes merge with 'Falco' changing into 'Mozart' in a quick montage. The video not only performs a literal dislocation of identity, but also conveys a 'touristy' message that if you want to imagine something from the past, think about something similar from the present. On this occasion, if we want to imagine Mozart, we can think about Falco and, by the same token, Falco can be seen as a new Mozart. However, I believe that by trying so ostentatiously to show that the past is like the present, that Falco is Amadeus, the video renders the past as ultimately unattainable. It should be mentioned that this effect is in part a function of the use of music video. As Carol Vernallis observes, 'music videos avoid Aristotelian narratives. The viewer takes part in the video's unfolding and notes the shifts of activity, affect, and time, but cannot fill in the context' (Vernallis 2004: 13). However, we observe no desire on the part of the authors of this video to make up for this limitation of the medium; on the contrary, they use it to its extreme to obliterate the difference between the past and the present and, by the same token, suggest that the past is unattainable.

If the subject of *Rock Me Amadeus* is the impossibility of reaching the Austria of the past other than by imagining it as a version of the present, the theme of *Vienna Calling* is the impossibility of reaching it from abroad. But this is not obvious from its title. It reads like a tourist slogan, referring to the city persuading the travellers to pay it a visit and admire its treasures. Vienna is among the most sung about cities in Europe, particularly during the interwar period. Especially iconic is *Vienna, City of My Dreams* (1914), popularised by the Austrian tenor, Richard Tauber (Gibson and Connell 2005: 20), which includes such passage:

> Heaven, Vienna mine
> I'm in the spell of your charms divine
> Dressed like a queen, with life so gay
> You are the love of my heart today

Vienna, City of My Dreams describes Vienna as if it was a woman, wooing a man with her divine beauty. Inevitably, by naming a song *Vienna Calling*, Falco evoked such a representation of his city. At the same time, it refers to a typical situation before the advent of mobile phones, when a telephone operator informed those waiting to be connected to distant places that their awaited conversation is about to happen. Of course, those most eager to phone Vienna were expatriates. Such a situation is presented in the lyrics of the song, which mixes the simulated voice of the telephone operator with a discussion about young women, who left Vienna to live in distant places such as Tokyo, Rio, and Toronto. The lyrics mourn the loss of authenticity due to travelling, as indicated by the fact that Vienna's natives change their names when they go abroad, and become metaphorically homeless. We can guess that among those homeless travellers is Falco himself, as indicated by mentioning 'Tucson, Arizona', a place where the singer shot a video to *Hoch wie nie* from the *Junge Roemer* album. However, there is a clash between the mournful lyrics and the cheerful disco music, suggesting that travelling and living abroad might in fact be pleasant.

The original video for this song is also ambiguous in its attitude towards the position of not being at home. It shows people dancing carelessly and phoning incessantly. Both activities render them dislocated from their everyday lives, yet not really transported to a different place, but rather suspended in no-place. Falco himself is a fragmented and even a spectral presence. First he is shown as an image adorning a telephone, then his walking and dancing is edited so rapidly that we get a sense of his discontinuous presence, as if he was magically transported from one place to another, but without even touching the ground. Although he dances against the background of other dancers, he seems oblivious to their presence and eventually they disappear, as if they were merely imagined. In this way, however, they also render Falco as unreal.

In the reworked version of *Vienna Calling,* assembled after Falco's death by the regular producers of his videos, DoRo, the anguish of being away from home is played down. A short sample from Johann Strauss's *The Blue Danube Waltz,* the most popular Austrian melody, begins the song, to give way to the sounds of telephone dialling and then the 'main' song. Most of the video is filled with the images of a handful of iconic Vienna locations, 'gems' and the manner of presenting them is typical for a tourist film or even a collection of postcards. Images are often distorted, to fit a specific object, for example a monument in a frame and a tilted camera is used to present it from a flattering perspective, as well as account for the way tourists photograph 'gem objects', again a technique discussed by Ning Wang (Wang 2000: 163–165).

No unpleasant image is shown which might disrupt the delight of a tourist; as if it was an illustration to *Vienna, City of My Dreams.* In the montage of different locations, as well as dividing the screen into two and on occasion even five parts, their historical context is lost: the whole of Vienna is reduced to places where tourists dwell, according to the same rule that Mozart's life was reduced to Vienna in *Rock Me Amadeus.* Falco in a dark suit, a tie, and orange shirt conveys a timeless image of a self-satisfied gigolo (although such an image is somewhat undermined by him wearing tennis shoes – a nod to the 1980s fashion).

Through the use of montage, in particular splitting the screen so he appears in the corner, he comes across as a guide to his beloved city, rather than somebody who left it to become a homeless tourist. The heterotopian content of the song is thus 'spirited away' to give way to an unproblematic touristy representation with 'Falco's ghost' as a guide to his city, no doubt prompted by commercial expectations that touristy images fare better than heterotopian ones.

The third song from *Falco 3* which engages overtly with tourism is *America*. Its protagonist moves between two locations and cultures: American and Austrian. It begins by pronouncing that the culture of a certain or every country is shaped by popular music; music provided it with a mythical beginning, as conveyed by using a slang version of the word 'Geschichte' (history or story) in this passage:

> Oiso,
> Die G'schicht is a jene,
> Des was a jeder
> Denn es liegt doch auf der Hand
> Es war mit Bock und Roll Musik
> Net immer leicht in diesem Land
>
> (The story is the following
> As everybody knows
> Because it is obvious
> It was never easy
> With Bock and Roll music
> In this country)

We can assume that this country is America, but the fact that it is not mentioned might suggest that the described situation is universal: the history of every country begins with 'pop and roll', although not in the actual, but metaphorical sense. There was a history of America before these styles were invented, but for an average American or a tourist visiting this country this (pre)history does not matter – we get in contact with it through listening to its music. However, after offering such an idealised description of music, as being the root and backbone of a national history, the song offers its less flattering description: as merely a commodity. Falco refers to his own popularity in America, mentioning that he is met with the words 'Falco you are wonderful', but claims (and rightly so), that this popularity is not translated into his listeners knowing him in any deeper sense, but merely buying his records and offering him cheap compliments. However, he repays this compliment and favour with the same superficial attitude, by drinking whisky. In this way he suggests that a record or music in general is, like alcohol, for quick and intoxicating consumption.

Later the song describes an American in Vienna, as revealing two cardinal sins of the tourist: vulgarity and superficiality. The American searches for typical Viennese attractions, epitomised by a Viennese schnitzel, and engages in excess, looking for a prostitute. His behaviour smacks of colonialism, not unlike that of the Germans travelling to Egypt, as described in *Tut-Ench-Amon* and is met with the same reaction: an attempt to exploit the naïveté of a tourist. The 'clever Viennese', not unlike 'a black boy on the corner, stalking like a vulture money on his mind' from the earlier song, offer the naïve tourists fake products at inflated prices as suggested by such a dialogue:

> 'I would like to have that wonderful Wiener Schnitzel'
> Würstchenhändler: 'Geh gib erm 100 dag Polnische
> in an Wachauer'

'Polnische in an Wachauer', a description of a type of sausage stands as a substitute for the exquisite Viennese dish, but also reflects the fact that Vienna is full of migrants, including people from the Slavic countries, who adopt Austrian identity for the perceived advantage of being seen as natives, such as working as guides for the naïve tourists. In the song, there is no real America or Vienna – tourism, understood as a form of commerce, reduces everything to fake images, things which pretend to be other things and are detextualised. Such representation brings to mind the previously quoted Debord's description of tourism as a by-product of the circulation of commodities, a chance to go and see what has been made trite. Yet, unlike *Hinter uns die Sintflut* or even *Tut-Ench-Amon* which state that the narrator of this song adopted a critical distance towards the observed reality, the protagonist of *America* is mentally outside it. He presents himself as a typical tourist, happily engaging in trite, commercial and cultural exchanges, which can be seen as a metaphor of Falco's overall attitude to his art, since he began his collaboration with the Bolland brothers and became a global star. The protagonist signals that there is no other reality than the trivial touristy reality as that described by Debord, but admits that he accepts this status quo because it provides material advantages.

The sense that tourism reduces experience to commercialised, fake, and trite pleasures is amplified by music. It draws, on the one hand, on Country and Western music and especially Bob Dylan and Bruce Springsteen, due the use of the mouth organ and the type of melody one associates with this genre. On the other hand, it is close to the works of Wolfgang Ambros (Lanz 2007: 149), one of the most popular Austrian pop musicians before Falco, the central figure of 'Austropop', who in his music embraced American culture, 'translating' it for an Austrian audience by emptying it of its edginess. It is also worth mentioning that in this song, Falco, apart from English, does not use standard German, but a Viennese dialect. Such a choice can be interpreted in contrasting ways: as an expression of a desire to relocate the American country or Western style to Austrian soil or a as gesture of resistance towards American culture.

Conclusions

My chapter attempted to present Falco as an artist who was both local and global, and engaged with different forms of travelling: tourism, colonial expansion, and time travel. I argued that his songs and videos produce an effect of heterotopia: an impossible space, in which different spaces and ontological orders mingle, but without producing a smooth, coherent space. Falco's songs rather point to the contemporary experience of time and space as fragmented.

Falco's songs also pose a question about the purpose and value of cultural translation – should we translate cultural artefacts and historical events for the sake of finding ourselves in a familiar world and being able to communicate with members of different cultures, or should we leave them unassimilated, forcing us to come to terms with the fissures in culture and history and accepting that we are locked in our times, national cultures, places, and languages. This question is never answered by Falco, but his grappling with it renders him as a fascinating phenomenon, deserving more attention than he has so far been granted.

Notes

1. Car crash deaths frequently add to celebrity status of the famous people and are typical sites of dark tourism (see Best 2013).
2. For example, many commentators on YouTube write that Falco made them interested in Vienna or encouraged them to even visit this city.
3. Daniel Jewesbury's discussion of the relationship between tourism and heterotopias on the example of London Bridge in Arizona Desert, is closest to my own

take on tourism and hetereotopia, as he points to the fact that the difference between touristy and heterotopian objects is a difference in representation.

4. Hans Albers (1891–1960) was one of the most famous German actors of the 1930s and 1940s. Falco was interested in his work and persona and at some stage wanted to make a record with covers of Albers' songs.

Works cited

Albers, Patricia C. and William R. James (1988). 'Travel photography: A methodological approach', *Annals of Tourism Research*, 15, pp. 134–158.

Best, Gary (2013). 'Dark detours: Celebrity car crash deaths and trajectories of place', in Leanne White and Elspeth Frew (eds), *Dark Tourism and Place Identity: Managing and Interpreting Dark Places* (London: Routledge), pp. 202–216.

Bohlman, Philip V. (2002). 'Jüdische Lebenswelten zwischen Utopie und Heterotopie, jüdische Musik zwischen Schtetl und Ghetto', *Lied und populäre Kultur*, 47, pp. 29–57.

Bolterauer, Alice (2010). '"Lost in translation" – Zur Funktion des Mehrsprachigen Zitats bei Hans Hölzel alias Falco', in Christian Ide Hintze (ed.), *Falco's Many Languages* (Vienna: Residenz Verlag), pp. 85–100.

Bork, Horst (2009). *Falco: Die Wahrheit* (Berlin: Schwarzkopf & Schwarzkopf).

Crang, Mike and Nigel Thrift (2000). 'Introduction' to Mike Crang and Nigel Thrift (eds), *Thinking Space* (London: Routledge), pp. 1–30.

Debord, Guy (1994) [1967]. *The Society of the Spectacle*, trans. D. Nicholson-Smith (New York: Zone Books).

Ernst, Peter (2010). '"Falconisch" – Falco und seine Sprachverwendung', in Christian Ide Hintze (ed.), *Falco's Many Languages* (Vienna: Residenz Verlag), pp. 125–135.

Foucault, Michel (1998) [1986]. 'Of other spaces', trans. Jay Miskowiec, in Nicholas Mirzoff (ed.), *The Visual Culture Reader*, sec. edition (London: Routledge), pp. 229–236.

Foucault, Michel (2002) [1966]. *The Order of Things: An Archeology of the Human Sciences* (London: Routledge Classics).

Gibson, Chris and John Connell (2005). *Music and Tourism: On the Road Again* (Clevedon: Channel View).

Harvey, David (1990). *The Condition of Postmodernity* (Oxford: Blackwell).

Hetherington, Kevin (1997). *The Badlands of Modernity: Heterotopia & Social Ordering* (London and New York: Routledge).

Jewesbury, Daniel (2003). 'Tourist: Pioneer: Hybrid: London Bridge, the Mirage in the Arizona Desert', in David Crouch and Nina Lűbbren (eds), *Visual Culture and Tourism* (Oxford: Berg), pp. 223–239.

Kilvert, Francis (1971). *Kilvert's Diary: Selection from the Diary of the Rev. Francis Kilvert, January 1870–19 August 1871*, vol. 1 (London: Jonathan Cape).

Kun, Josh (2005). *Audiotopia: Music, Race, and America* (Berkeley: University of California Press).

Lanz, Peter (2007). *Falco: Die Biographie* (Wien: Ueberreuter).

Mazierska, Ewa (2014). *Falco and Beyond: Neo Nothing Post of All* (Sheffield: Equinox).

Potter, Russell A. (1995). *Spectacular Vernaculars: Hip-Hop and the Politics of Postmodernism* (New York: State University of New York).

Shields, Rob (1991). *Places on the Margin* (London and New York: Routledge).
Straw, Will (1993). 'Popular music and postmodernism in the 1980s', in Simon Frith, Andrew Goodwin and Lawrence Grossberg (eds), *Sound and Vision: The Music Video Reader* (London: Routledge), pp. 3–21.
Toussaint, Stéphanie and Alain Decrop (2013). 'The Père-Lachaise Cemetery: Between dark tourism and heterotopic consumption', in Leanne White and Elspeth Frew (eds), *Dark Tourism and Place Identity: Managing and Interpreting Dark Places* (London: Routledge), pp. 13–27.
Urry, John and Jonas Larsen (2011). *The Tourist Gaze 3.0* (London: Sage).
Vernallis, Carol (2004). *Experiencing Music Video: Aesthetics and Cultural Context* (New York: Columbia University Press).
Wang, Ning (2000). *Tourism and Modernity: A Sociological Analysis* (Amsterdam: Pergamom).

9

In Praise of Authenticity? Atmosphere, Song, and Southern States of Mind in *Searching for the Wrong-eyed Jesus*

Nick Hodgin

'It so happens' writes Michael O'Brien, 'that a disproportionate amount of American popular culture [...] is southern. Jazz, blues, rhythm and blues, rock music, country and western, much in those genres is southern or part of a southern cultural diaspora'. He goes on to mention the omnipresence of depictions of the South in film and television, the influence of Southern literature before claiming that 'to know the South is indispensable to understanding America' (O'Brien 2007: 11). It was not always thus: James Cobb argues that the late arrival of Southern music – especially country music – to the rest of America occurred at a time (1970s) when the nation was adjusting to the 'twin shock of defeat and disillusionment previously only associated with the experience and heritage of the Southern states' (Cobb 1999: 78). It is worth noting that Southern music – hillbilly music – was not simply unknown previously, it was actively reviled for it was seen as the noise made by the primitive half of the US. If New England was understood as the 'genesis and crystallization of "American civilization"', argues Larry Griffin (2006: 7), then the South was 'America's opposite, its negative image, its evil twin'.

Ostensibly a geographical reference, the South is of course much more than an issue of cartography. Just as Goethe once remarked of his homeland, then a collage of kingdoms and dukedoms and principalities, 'Germany? Yes, but where is it?', so too the South, is more an idea than it is a region defined by uniformly agreed borders. Its cultural diversity is a symptom of its vague geography – from the bayous of Louisiana to the Texan plains, to the Tennessee mountains. Thus there are different versions of the South fine-tuned by various prefixes – old South, Deep South, antebellum South, post-reconstruction South, and so on. This region has entered the cultural imaginary even among people far removed from its geographical and cultural contexts. It is not unreasonable to say that most people in the UK would be able to provide a number of reference points if asked to say what they associate with the American South. Those of a certain age and education would perhaps cite Flannery O'Connor or Tennessee Williams or Truman Capote but some would know to say John Grisham or Cormac McCarthy, maybe Bill Clinton and Jimmy Carter before him. The South's elegance (the Old South), meanwhile, is associated with moral and physical corruption in the form of political or physical abuse or general material decay. More likely associations would stem from popular culture references. *Deliverance* (John Boorman 1972) is often mentioned as reference to a backwoods primitivism associated with the rural South (and all that that apparently entails); Cletus, the slack jawed halfwit with umpteen children in the *Simpsons* might be another, and there has been a host of television series – often made by people not from the region – either set in (but rarely about) the South (*Lizard Lick Towing NC*; *Deep South Paranormal Research*), or making good use of bad Southern characters which have further set our understanding of the region according to some fundamentally negative stereotypes and ideas – including low IQ, incest, white supremacism, religious extremism, criminality. And it begins early. Children's television programmes frequently make use of the South: in an episode of *Ben 10*, aliens who are intent on destroying the earth are coded as Southerners. All the markers are present: the Southern drawl, the gap toothed grin, the slow wits, and dungarees. The perennially popular *Scooby Doo* is one of many programmes to employ negative markers of the South, with occasional cameos by 'Scooby Dum', a moronic, buck-toothed hound with a distinctive Southern pronunciation. Media representations have, then, done much to shape our impression of the South, a region much imagined but seldom visited (geographically, Orlando and Miami are the best known tourist destinations but considered distant from the South in all other senses).

For good or for bad, what is clear is that the South endures (to riff on the title of John Shelton Reed's landmark text, *The Enduring South*) as a region that is distinct from others in the Union (Reed 1972). Asked what they think when they think of the American North or the East Coast or the West Coast and the chances are that most Europeans will focus simply on individual cities (New York, Boston, Chicago, Seattle), perhaps California state, while the mid-west is likely to be characterised as a void between better imagined coordinates or simply flat and empty. The interest in the South has more than endured. It has, as Dubey notes, proliferated: 'since the mid-1970s, U.S. historians, sociologists, novelists, literary critics, and cultural commentators seem to have become obsessed with the South, reviving the enduring debate about what makes the region distinct from the rest of the nation' (Dubey 2002: 351). This chapter examines this distinctiveness, looking at its mediation through music and film as presented in a collaborative documentary film by a British director and local musician, paying particular attention to the issue of authenticity. Authenticity has, as I show, long been a concern for musicians especially where these are seen as being representative of a particular place and culture and regardless of the accuracy of that association. Music functions as a signifier for the authenticity of place, an expression of the authentic culture that we hope to explore and experience. Understanding that authenticity is as much about performance and reception, that it is a construct and a quality ascribed, proves to be particularly relevant when reflecting on the ways in which the music of the South has been performed, promoted, and consumed.

Searching for the Wrong-eyed Jesus: an impossible quest?

In *Searching for the Wrong-eyed Jesus* (2003), Andrew Douglas's award-winning documentary film made for Arena, the BBC's influential arts series, musician Jim White travels across the rural South searching, as White puts it, for the 'gold tooth in God's crooked smile' or, in the British director's more prosaic words, 'the soul of the South'. The objective is, as both readily concede, impossible, but this does not prevent them from trying to establish what it is about the South that has so mesmerized artists, particularly musicians. The director's interest in exploring the region's musical legacy was partly prompted by listening to Jim White's debut album *The Story of How I Shouted 'Wrong Eyed Jesus'* (Luaka Bop, 1997) and he was able to convince the American singer-songwriter to act as a kind of tour guide of the region (yet another role that the one-time model, former professional surfer, and erstwhile NYC cab driver could add to his CV).

The film travels across several Southern states in search of an essential Southerness calling in at juke joints, burger bars, Pentecostal churches, and a prison, and features dialogues as well as monologues about the South, often poetically rendered, occasionally surreal or inflected with religious references, which are intended to provide further insight into what the South really means. This involves a range of people of a kind seldom seen in mainstream film and television: gap toothed former drug addicts, tattooed trailer park residents, fervent preachers, shaven headed prison inmates, many of whom represent the poor and the disenfranchised across several generations, from barefooted kids interested in the film team to elderly banjo players and wizened nuns. *Searching for the Wrong-eyed Jesus* resists simple categorisation and is equal parts travelogue, film essay, and musical voyage. Though conscious of the impossibility of the quest, the filmmaker and guide do, however, seek to provide some insight into the real South and they do so by accentuating the region's authenticity.

That authenticity should serve as a line of enquiry for this film in particular is not surprising given that it is frequently attributed to the South, often framing discussions of the region's music, especially where music is the cultural expression of the poor. Authenticity is of course a complex, polysemic concept with a long history of critical and theoretical discourse especially in the twentieth century and since. Of relevance to this chapter is the postmodern interest in authenticity and simulation, between original and copy, explored by Eco and by Baudrillard, both of whom focus in particular on North American culture as one characterised by a search for the authentic and by the proliferation of copies, fakes, and reproductions, which in Eco's words provide 'reassurance through imitation' (Eco 1998: 57). Where for Eco the desire for the real can be attained only by fabricating 'the absolute fake' (Eco 1976: 42), (mis)understood to be more real than that which it imitates, Baudrillard sees no remnants of the original, proposing instead a world defined by its absence, a hyperreality of 'perpetual simulation' (Baudrillard 1989: 76), a more final reading than Eco for whom hyperreality is characterised by the 'confusion between copy and original' (Eco 1976: 42). A similar line of thinking informs Barker's and Taylor's (2007) analysis of authenticity discourse in popular music and Mark Fisher's thoughts on contemporary pop culture. In *Ghosts of My Life*, the latter reflects on the 'peculiar temporality' (Fisher 2013: 10) against which some bands evolve, noting the reliance on past musical scenes in order to achieve some integrity through citation of past genres and styles. Performed anachronism is often a self-conscious undertaking, one that serves to differentiate the bands from the perceived vacuity of contemporary music and in this they are the equivalent of what Eco calls the 'offspring of the unhappy awareness of a present without a depth' (Eco 1976: 31).

Authenticity and place: armchair tourists and cultural explorations

In Douglas's film, authenticity is partly achieved through the interviews with some of the region's carefully selected natives, whether bikers or inmates or preachers; elsewhere it is suggested via symbols and material artefacts that resonate with the aura of authenticity. The instruments played (vintage guitars, dulcimers, washboards, saws, homemade banjos, fiddles) connote authenticity, providing both a tangible as well as conceptual connection with the past. The spatial arrangement, too, augments the common view of the South as unpolished, natural (adjectives that are applied as much to the region's local music as they are to the place): reality is staged and the stages in question include a crumbling building, a ramshackle shed, a junkyard, an empty houseboat, all of which are suggestive of an imagined past, settings for the musicians-as-revenants whose music and intonations channel past spirits, an example perhaps of what Greil Marcus in his discussion of Dylan's and the Band's *Basement Tapes* (Columbia, 1975) calls 'a vague tourism of specters of a foreign country' (Marcus 1997: 86). That foreign country, which for most is simultaneously familiar and unknown, is the South – or at least one version thereof. What is interesting is that none of these markers and signs is especially surprising and that is because of a compositional understanding of the South, a problematically imagined cultural community sustained, as previously mentioned, by a variety of media able to produce images that register with long embedded ideas about the South. But Southern exceptionalism is a symptom of homogenising and self-mythologising as much as it is of ascription and the 'tourist gaze': the Jack Daniels adverts, for example, in which earthy characters in dungarees in barns move between whisky stills in sepia-toned images accompanied by a fiddle and banjo soundtrack is an example of the 'closed self-perpetuating system of illusions which provide tourists with the basis for selecting and evaluating potential places to visit' discussed by Urry and Larsen (2011: 7).

The audience-as-tourist has no agency, no means of deciding on the route to be taken or places to be visited and must rely on the filmmaker in the company of his local escort. But White is not an altogether reliable guide. His narration is subjective, anecdotal, at times rehearsed. Though often offbeat, humorous, the astute sociological observation is important for it provides context to images that might otherwise exoticise the subjects as examples of the 'southern weird' (a term applied to White's own music by reviewers), another popular conception of the South (Gill 2004). While weird denotes something odd, grotesque, perhaps inexplicable, the manifestations of the latter speak more to the economic realities of the South and the culture thus produced than they do of a regional exceptionalism or of eccentricity as an inveterate characteristic of the South's population. His attitude to his role remains ambivalent for he is caught between the desire to demystify the South and to preserve its mystery. This is partly because White is aware of the South's untranslatability; at one point he describes being overwhelmed by the South's immensity. Nevertheless, his insider knowledge of cultural practices – the religious circumstances in particular – imbues the film with a degree of authenticity, albeit one that is also performed: choosing the right car for the road trip, one of the first decisions on his itinerary, is an acknowledgement of needing to look the part, an example of 'dressing up' which is a frequent practice in simulating authenticity. Knudsen and Waade stress the importance of performative authenticity, noting that 'whether one is a performing body or a city/region/country, it is possible to "authenticate" sites, sights, places and to enhance the tourist's/traveller's understanding and their sense of intimacy, self-reflection and feelings towards their surroundings' (Knudsen and Waade 2010: 2). The focus here is on the tourist engagement with place and space but not irrelevant to the relationship that develops between White the narrator-guide and us, the armchair tourists.

While the film is described, tongue in cheek, as being something between a religious mission and ethnographic quest, the film is really an exploration of the South through music, if not exactly the music of the South. And the purpose of such an enterprise is to gain some insight into what the South really means, to satisfy audience's curiosity about the region. Such a quest is driven by the same impulse that guides travellers and tourists to venture into other cultures and therein to connect or witness the authentic in its natural context, to discover the real, and this impulse is one invoked both directly and indirectly in Douglas's film. A motivating factor for tourists' search for authenticity is, as Dean MacCannell noted in is his seminal essay, the 'shallowness of their lives and inauthenticity of their experience' (MacCannell 1973: 589–590). Drawing on Erving Goffmann's dramaturgical model and his notion of front and back regions, Maccanell applies this to the tourist's search (and we may apply this to both armchair and mobile varieties) for that which is out of the way, remote, unspoilt – tropes so common within traveller discourse, from broadsheet travel supplements' 'off the beaten track' recommendations to multiple edition travel guides, that the terms anticipate their own incongruity. The appeal of those back regions is one he explains thus: 'Just having a back region generates the belief that there is something more than meets the eye; even where no secrets are actually kept, back regions are still the places where it is popularly believed the secrets are' (ibid.: 591). Back regions afford us a connection with that which is perceived to be more real, more intimate, to 'see behind the others' mere performances, to perceive and accept the others for what they really are' (ibid.: 592). The intimacy to which Maccennell refers is, as he acknowledges, an issue that dogs ethnographic research and it is one that again raises the issue of authenticity since the search for authenticity is one that is frequently felt to be possible if only one can somehow connect with local people. This sense of communion with locals is part of a desire to transform them from tourist-objects to host subjects, to reconfigure the inevitable bottom-down vertical connection in a desire to establish some horizontal engagement, however fleeting that may be. We might note that such horizontal engagement is what usually makes witnessing a live performance the pinnacle of contact experience with musicians (Connell and Gibson 2003: 37) – and the more intimate the venue, the more authentic the experience seems: witness the cosy mise-en-scène arranged for acoustic sets performed by bands more accustomed to stadium performances.

White, the local interloper, goes some way to facilitating the desired connection, offering the film team/audience a personal tour through the region. The journey is a pilgrimage of sorts, complete with a small statue of Christ leaning out of the boot of his car, narrated and led by a long-term resident whose experiences and biography lead us to confer on him the status of authentic guide. His knowledge of the area certainly makes for some intriguing (back region) locations, including back roads, junk yards, trailer parks and porches, and masks the film as one by a British filmmaker. His connections via family and via his music also enable a series of encounters with various musicians, whose music attests to different resonances that the South has had on dissimilar people and different musical genres and who are often – and problematically – presented as representatives of a Southern tradition.

As important to the film as the music and the people is the landscape, filmed in Florida, Louisiana, West Virginia, which is beautifully presented (Douglas's previous work as director of high concept commercials is apposite). A slow camera pan reveals trees webbed with lichen; we see the waterways lined with vegetation through which the sun filters, dark knotted forests, hills, sunsets. The presence of the region's inhabitants is not occluded but the film tends towards a depiction of the landscape that privileges the natural over the man-made and in which decay and pastness, tropes common to many of the photographic representations of the Deep South, are foregrounded – weathered shacks, cars rusting in wrecking yards, dilapidated properties. The film takes pleasure in the ruins which are testament to times past and people gone. Svetlana Boym has sought to distinguish between ruinophilia and nostalgia, arguing that 'ruinophilia is less afflicted by the personal story; it is not a longing for home or for identity but more of a material and visceral experience of the irreversibility of time that comes together with care for the world' (Boym 2008). While Boym is right to uncouple the two, for ruinophilia and nostalgia need not be mutually dependent, nostalgia is more often than not ruinophilia's emotional undercurrent if not its logical synonym. Ruins are usually suggestive of human failure. They may signify economic neglect, architectural hubris, the decline of civilizations, or man's failure to withstand natural forces, but the fascination with them may be less a symptom of a contemporary melancholic condition; instead, as Boym argues, ruins can be 'sites for a new exploration and production of meanings' (ibid.). The ruins, debris, and artefacts seen throughout the film are as much about the present and future as they are about the past. We are given little indication that the region has a future; if it does, it is one that moves forward but remains turned to the past. This retrospective turn is neither unproductive nor devoid of meaning, however. Indeed, the past provides coordinates for present and future exploration – and this is certainly true of the film's music.

The real sounds? Music and authenticity

According to White, music is central to our understanding of the South. The connection between music and place, where each informs the other, has become commonplace in the characterisation of certain regions and countries. Travelogues, music magazines, guidebooks, and commercial brochures typically argue that only by understanding and appreciating particular musical traditions can one expect to gain any purchase on the wider local culture. Such cultural exploration has resulted in the musical culture of certain locations being elevated above others; the significance accorded it by outsiders may even be disproportionate to its relevance within the community of origin, a point repeatedly made by Barker and Taylor (2007: 30–99) in their consideration of the reception and circulation of the blues. Among those places whose musical traditions have been ascribed a significance by outsiders and whose financial potential in the form of touristic patronage has in turn been recognised by those inside the tradition are countries such Cuba, Mali, Ireland, regions such as Andalusia, and cities such as New Orleans and Seattle ('birthplace of jazz' and 'home of grunge' respectively) – 'distinctive national styles [...] marked by strategic essentialism, marketing and local boosterism' (Connell and Gibson 2003: 124). One of the consequences of this external categorisation is that in order to satisfy market demand, the original object of curiosity (the people and their modes of behaviour – and this applies as much to other forms of essential expression whether cuisine, dance, or music) spawns ever more simulations, each claiming to be the authentic article. Whether authentic or not, certain articles become synonymous with the experience of place by virtue of association. The HBO series *Treme* (2010–2013), which focuses on New Orleans in the aftermath of Hurricane Katrina, emphasises and satirises in some episodes the difference between the authentic and the simulated, revealing the shallowness of outsiders' knowledge of music and indigenous musicians' willingness to capitalise on such ignorance. Thus the New Orleans musicians' commitment to their music is never in doubt, but musical integrity and financial need does not preclude their participation in better paid gigs, usually for tourists looking to experience the 'real' New Orleans and usually with a set list of standards we understand the musicians would never consider playing privately (notably *When the Saints Go Marching In*). Analysis of certain musical genres is similarly fixated on place, on the environment as both determining factor and symptom of musical development, so just as musical genres in Brazil resonate with assumptions concerning the Brazilians' apparent sexiness and laid-back attitude to life (ideas that certainly informed the related coverage of the 2014 World Cup), so commentators reflecting on musical genres in Britain see a causal link between the Midlands' heavy industry and the region's heavy rock/metal origins or suggest a correspondence between the imagined dystopian, becoming-post-industrial Sheffield environment of the late 1970s and early 1980s and the city's stark electro sounds (as suggested in the BBC documentaries *Metal Britannia* (Chris Rodley, 2010) and *Synth Britannia* (Ben Walley, 2009), respectively).

Musicians and music scholars have long been preoccupied with what constitutes the authentic, whether in discussions about song styles (in Britain in the late nineteenth century rural folk songs were eulogised while modern music hall songs were vilified), the kinds of instruments played, or the proprietorial concerns that surround certain musical genres. The scepticism towards perceived posturing is evident for example in Benjamin Filene's discussion of the folk music revival and the Coen brothers' film *O Brother Where Art Thou?* (1996), when he states, 'I'm not a purist: I'm not griping about the fact that the performers benefiting from the revival are long-time commercial popularizers like Emmylou Harris or a Californian neo-billy like Gillian Welch' (Filene 2004: 50). That ownership, the question of musical expression as expression and component of identity has long been a factor in the discussion of certain musical forms – and an abiding concern in popular music discourse: can millionaire white guitarists really play the blues? Is German rap not a contradiction in terms? Can there be such a thing as British trad jazz? The interest in certain musical traditions has (partly) been sustained because of their perceived authenticity. As Filene notes, 'the idea of returning to American roots – so central to the 1930s revivalists' mission – is again a legitimating rite of passage for younger artists and a way for older artists to reinvigorate themselves' (ibid.: 53). Explaining the foreign appeal of blues in Britain, Allan Moore suggests that the 'desire to appropriate the "unmediated expression" which was thought to be the preserve of the country blues style, entailing an unquestioned assumption that African Americans in the Southern USA were somehow more "natural" beings than white, college-educated Londoners' and something similar has happened with the excavation of musical traditions of the South – folk, country, bluegrass, and other so-called old-time music' (Moore 2002: 215).The commingling of the South and authenticity is a feature on which musicologists and musicians routinely focus and which has not infrequently resulted in an idealising of the South's backwardness, which tends to avoid socioeconomic circumstances focusing instead on a romanticised and originary state. This is by no means a recent development. In *American Ballads and Folk Songs,* published in the 1930s, John Lomax wrote 'A life of isolation, without books or newspapers or telephone or radio, breeds songs and ballads' (Lomax et al. 1934: xxvii). Lomax, like that other great collector of vernacular song, Harry Smith, recognised that the music's origin was not incidental to (mostly urban) audiences and markets, hence his sleeve note emphasis on 'authentic folk songs performed by genuine folk singers' and the implication that the poorer the person, the better the music (quoted in Allen 2010: 28). The connection between economic hardship and rich musicianship has long been the preferred logic of critics and fans and characterised the discourse surrounding the blues genre in particular. This focus on authenticity, however, serves as a kind of creative straitjacket, even if the claim of authenticity is often predicated on simulations of the authentic: the criticism of Bob Dylan for abandoning his roots assumed an authentic background that Robert Zimmerman had fabricated, a fact that apparently was less objectionable to the scene's 'cult of authenticity' (Yaffe 2009: 21). Interestingly, Feline uses the same term in his discussion of contemporary focus on the authentic, arguing that the 'cult of authenticity not only stymies performers [...] it tangles up listeners as well' (Filene 2004: 65). In their discussion of the blues scene and its self-appointed aficionados, Barker and Taylor note how Lonnie Johnson's decision to sing a Fred Astaire song at a folk music concert 'confounded expectations' of blues and folk fans and that he was never invited to perform there again (Barker and Taylor: 84). Keith Richards, who is self-aware (and unrepentant) of his and his bandmates' interpretation and imitation of their blues favourites remembers the period as one in which 'ludicrous discussions about authenticity' (Richards 2010: 82) were routine and on occasion resulted in violent confrontation. Audiences' preferred version of Dylan as committed folk troubadour and resistance to his transformation in 1965 from pre-modern acoustic to an assumed inauthentic electric sound was not unprecedented, nor would it be the only time that the singer's musical explorations would disappoint fans. His country record, *Nashville Skyline* (1969), likewise divided critics. Big Bill Broonzy, too, had some years earlier been celebrated as an authentic exponent of folk blues, a genre associated with the South, and a role that the urbane, Chicago-based Broonzy was astute enough to recognise would be far more profitable, especially among well-paying and enthusiastic white European audiences seeking to connect with the authentic blues, a desire that Broonzy duly fulfilled, even trading in his dapper city suits for modest rustic garments (Barker and Taylor 2007: 186–187).

Claims of authenticity characterise much of Southern culture or rather the branding of Southern culture: 'authentic' is an adjective that is attached to or implied everywhere in the marketing of the South, whether in cookbooks of regional cuisine or local recordings or radio station programmes or tourist guides ('experience the real South', 'see the true South') all of which amount to an exceptionalising of the region and its culture. But the authentic, as Moore reminds us, is 'ascribed, not inscribed' (210) a point echoed by Lisa C. Roberts: 'The experience of authenticity [...] is based less on an inherent quality than on a sign imposed from without' (Roberts 1997: 101). It is possible that native Southerners would refer to themselves or their various modes of living as 'authentic' but it is unlikely, a fact often overlooked by the arbiters of what constitutes the authentic.

Interspersed throughout the film are interviews with locals (including a monologue on place and memory by writer Harry Crews) and musical performances by White himself, by Johnny Dowd, the Handsome

Family, 16 Horsepower's David Eugene Edwards, octogenarian banjo player Lee Sexton, Melissa Swingle (from Trailer Bride) and, bizarrely perhaps, former New York Doll frontman, David Johansen. There is a near constant musical soundtrack (both diegetic and non-diegetic) including church songs, music played on the radio, and three women (including White's ex-mother in law) singing harmony in a fast-food restaurant. These are not introduced; nor is there any commentary. Scenes simply expand to include them: the camera moves from the action to a performance; other performances are simply set between scenes. But there is no narrative logic that guides their organisation (Figure 9.1).

Why is music so prominent a feature of a film about the American South commissioned by the BBC's 'Arena' series? White argues that wherever you go in the South you will find people playing music and singing songs. Southern culture is, according to White, defined by two distinct paths: that which leads to God and that which leads away from God, and music, he argues, accompanies both of the routes taken. If we assume, then, that a film about the South must include music, then the question perhaps should be why are these particular musicians in the film? Eugene Edwards is from Colorado; Rennie Sparks, the Handsome Family's lyricist is from Long Island; Johnny Dowd, though born in the South has long been a resident of New York state; and if

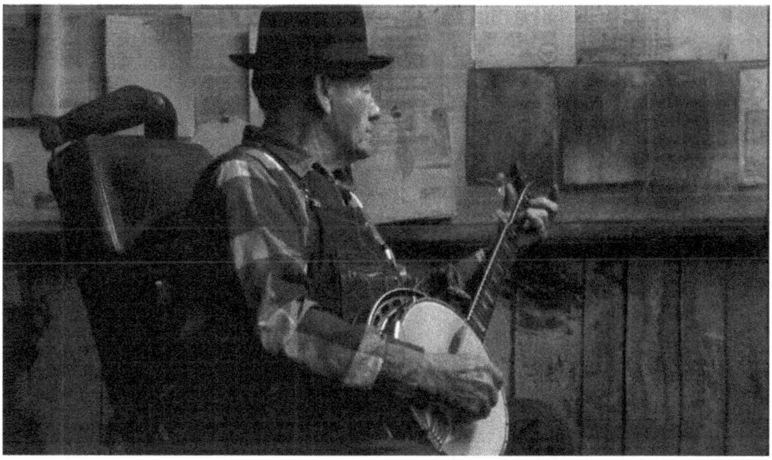

Figure 9.1 Authentic settings, authentic sounds? Lee Sexton performing
Source: Bill Douglas, 2004.

ever there were a musical and visual antonym for what constitutes the typical South, Staten Island born David Johansen, lead singer of the New York Dolls, would surely be a contender. Relocating (some of) these musicians to the South for the film in order to demonstrate the significance of Southern music seems to undermine the film's original purpose – namely to grasp a sense of the South. But the film is not an ethnographic project and neither White nor Douglas claims to be undertaking genuine research into the region's traditions or vernacular song. These musicians' disconnection from the local people, whose region they are (in)directly representing is even foregrounded by the choice of venues. The motel room is which Johansen sings is suggestive of the visitor, the musical tourist; Edwards, plucking his banjo, wanders alone among the trees, while White is seen in his porch and at a petrol station and so on. These are performances not for local crowds but for imagined audiences.

What we infer from the film is the importance of musical heritage, though this might be more accurately described as an imagined tradition, one that fills an implied void. All of the musicians' musical choices, whether lyrics or genres or instruments played, acknowledge a debt to the South, a region whose culture has resisted absorption into homogenised mainstream culture and seemingly retained something authentic. The interest and fascination with the South has a history too complex to explore here but one explanation is that the romantic view of the South reflects the North's nostalgia for a pre-industrial antebellum period, which may compensate for the alienation typically associated with those urban, industrial centres in the North, stereotypes that exist 'despite the growing industrialisation, urbanization and national slant of the South' as Cecil K. Hutson commented more than 20 years ago (Hutson 1993: 56). It need not be viewed in terms of American continental culture: Nick Cave, an Australian whose debt to Southern music and literature is well known, has said of the context of his writing, 'These are romantic notions I've had as somebody who's never been to the Deep South. I don't see my songs as being Southern [...] It's just a mythological territory I've devised as a stage where a lot of my songs and this book [*And the Ass saw the Angel*] operate' (quoted in Smith 1990).[1] It is less the indigenous population than the outsiders, then, who have helped establish Southernness as a separate identity and culture (the resurgence of Cajun culture for example, was largely due to the interest in an authentic Francophone culture by visitors from Quebec, France, and Belgium).

Some of the musicians who appear in Douglas's film are inspired by the religious rhetoric of the South, others moved by the creative spirituality of Pentecostalism. Renée and Brett Sparks of the Handsome Family discuss 'the art that emerges from the clash between the sacred and the secular' and suggest that the 'Devil is alive in the South'. 'What is it about Southern music', asks Rennie Sparks, 'that it has a lot of blood and violence and fiery images?' before concluding that Johnny Cash and Jerry Lee Lewis could as easily have been killers or preachers as musicians. Others cite the writers of so-called Southern Gothic as literary influences (Eudora Welty, Faulkner, O'Connor); those who are from the South, such as octogenarian banjo player Lee Sexton, introduce their songs by explaining their origins, thus underlining these as cultural products of a specifically Southern tradition. Johansen's interpretation of *Last Kind Words*, Geeshie Wiley's 1920s blues song, meanwhile, is the reason for his appearance. In addition to his New York Dolls work and the eccentric Buster Poindexter lounge lizard alter ego, Johansen has a long interest in the blues and in recent years has interpreted several rare songs collected by ethnomusicologist, Harry Smith, whose name he has appropriated for his backing band.

Interpretation is key here. The musicians are not reproducing Southern songs but interpreting them, giving voice or melody to their experience or understanding of the South. Some offer a version of their source influence whose end result is more creative interpretation than faithful reproduction while others are more concerned with fidelity. Johansen, for example, manages to sound a lot like the African-American woman recorded by Smith, though I am conscious of Octavio Paz's comment that the world is increasingly presented through collated texts, each a variation on the one that preceded it so that what we have are 'translations of translations of translations' (Paz 1992: 154). This is a potentially upsetting thought for the disciples of Smith's recordings, which are often treated as if they were 'originals', unique fragments, echoes of persons and places lost to time. Others, meanwhile, are seeking somehow to translate the experience of the South (and though few would claim that theirs is the essential truth, most I think would say that it approaches the truth, even if it is a truth that is poeticised and glamourised). It might be more accurate to say that these artists seek to synthesise ideas and tropes relating to a Southern imaginary, its values and complex narratives, into music. Dowd's noirish songs ('There's been a murder here today. Bloodstains are on the wall') are inclined towards the (Southern) Gothic; the Handsome Family's songs, too, are dark narratives; White's texts are more lyrical (and often combine the fantastic and strange, a vision that presumably accounts for reviewers' use of the term 'southern weird'); Edwards, previously bandleader of the alt country group, 16 Horse Power, and now singer of a Christian band, sings *Wayfaring Stranger* while another standard, *Amazing Grace*, is played on a saw. These songs are presented as a means of understanding and communicating something of this Southern culture, though the target audience presumably is not indigenous Southerners. In this sense, they are producing something related to but not resembling the original and, in so doing, approximate Walter Benjamin's idea of translation as 'translation, instead of making itself resemble the meaning of the original, must lovingly, and in detail, fashion in its own language a counterpart to the original's mode of intention' (Benjamin 1997: 161).

The director uses the musicians to capture some essence of the South as he imagines it and, with guidance from White et al, to do this in a way that is comprehensible to outsiders but which inevitably involves conventional markers of the South and of Americana in general: the swamps and creeks, banjos and dungarees, a swinging gas station sign, breathless Pentecostal preachers, a Confederate flag as an unsuitable backdrop for the Handsome Family, etc. The musicians seek to extend the reality of the South beyond its native cultural confines even if, paradoxically, this means sustaining a mythical South that lacks necessary exposition. Such a knowing distance between source and copy and the willing suspension of disbelief on the part of the audience is not unusual: as Miles Orvell suggests, 'our own time might be called a culture of the factitious. We have a hunger for something like authenticity, but we are easily satisfied by an ersatz facsimile' (1989: xviii). The logic implied by 'something like authenticity' might be understood thus: they may not be the real thing, but they are close enough (a sentiment that reaches its fullest expression in the reception of tribute bands). This self-awareness has characterised a number of reviews discussing some of those artists who appear in *Searching for the Wrong-eyed Jesus*. One review of 16 Horsepower's music, for example, comments on the authenticity as performance, noting that 'David Eugene Edwards is a singularly possessed talent, arguably the most capable of his generation's peddlers of rural American doom, donning the authentic field garb of generations buried in miles of pastoral dust' (Fink 2003); another meanwhile claims that 'In a world where the term is bandied about far too liberally, Johnny Dowd is definitely "authentic", albeit with a healthy dose of sarcasm and a pleasing sense of the absurd'(Smith 2010).

The examples of musicians obsessively seeking to enhance their sound by literally channelling their music through authentic equipment are legion: Neil Young's instant recording in a 1947 'Voice-O-Graph' booth (*A Letter Home*, Third Man Records, 2014), which enabled him to age and retrofit a collection of cover songs, received mixed reviews, from those sceptical of the 'faux authenticity' (Sweetman 2014) to those for whom the antique equipment 'seems to give his singing its strangeness back' (Petridis 2014). Even more intriguing is the playful fabrication of authenticity by contemporary electronic artists who, in fashioning crackles and hiss on their digital recordings also seek to age their music but in ways that, in Fisher's words, 'make "out of joint time" audible' (Fisher: 144). The temporal disjuncture Fisher suggests is not of course unique to the self-conscious anachronisms nor is he the first to comment on such discontinuities. Discussing the Handsome Family's oeuvre, Asbjørn Grønstad notes that their sound, too, is 'anchored in space yet adrift in time' (Grønstad 2005). Nor is the desire to intensify authenticity through the use of obsolete or neglected equipment unique to musicians. A similar aesthetic decision is evident, for example, in Sally Mann's approach in her photographic work, titled *Deep South* (2005), which seeks to illuminate the real South but also to capture what she calls its 'radical light', an aim she pursued by using the nineteenth century collodion or 'wet plate' process, thus literally viewing the present through the lens of the past and, in the process, conferring on the South again the nostalgia and 'an almost tangible authenticity' with which it is associated.[2]

Conclusion

One might argue that the South is presented according to a universally understood visual syntax (by universal I mean Western) and therefore requires very little decoding. Indeed, we would probably struggle more with the film and music if they were to be presented otherwise. Imagine the difference in seeing the elderly Lee Sexton perform in one of the many Museums of Contemporary Art that can be found in the region. There is then a sense of generic expectation, the requirement that the representation of authentic culture conform not to what *is* authentic but what *is thought* to be authentic (and here one might think back to the staged reality of Flaherty's *Nanook of the North* (1922) or of Sitting Bull presented to audiences in the late nineteenth century wearing full Native American dress as part of Buffalo Bill's enormously popular travelling show) (Figure 9.2).

It is important to emphasise that White and Douglas do not conceal the fact that most of the musicians whom they have invited are

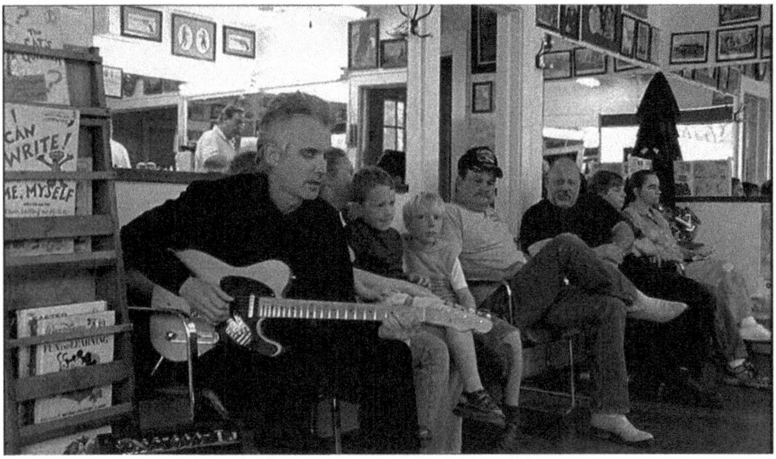

Figure 9.2 A southern audience tolerating Johnny Dowd
Source: Bill Douglas, 2004

not representative, or even part of, the local culture. As mentioned, certain scenes even highlight their difference: White plays one of his songs in a restaurant and manages to stimulate no interest from the locals at all. His disconnection from other Southerners is even emphasised when he removes himself from the bar, which, ironically, is the one location where locals are seen enjoying music, not a live performance of authentic Southern sounds or celebration of Southern music but generic rock played through the bar's speaker system. The performer's distance from the people is evident, too, when Johnny Dowd sings a duet in a barber's and is viewed with wry amusement by those waiting for a haircut. The Handsome Family appear in isolated locations: the houseboat, a roadside, inside a car, the only vehicle in an empty parking lot. At one point they are seen singing at the edge of a trailer park singing to no one but a puzzled young boy. Such scenes pre-empt any accusations of a 'staged authenticity' in the sense that McCannell means in his discussion of tourist spaces; if anything, they foreground these musicians as atypical, as non-representative. The indigenous Southerners show little interest or appear not to recognise themselves or the South in these representations. Indeed the only instance of locals engaging with live music is a scene shot inside a church where congregants respond enthusiastically to the preacher/musician (not something that is featured on the film's soundtrack CD). Missing from the representation, too, is any reference to the considerable African-American community, an absence that received comment in a number of reviews. *Variety*'s description of the film as 'a very European look at a quintessentially American milieu' (Felperin 2003) is self-evident; a more useful explanation is that offered by Michael Booth of the *Denver Post* who suggests that the recourse to stereotypes permits us 'to understand those cliches [sic] in a new way' (Booth 2005). Booth's reasoning recalls Ayer's statement that:

> The South has suffered from generalizations that trivialize it, whether those generalizations take the form of romantic and nostalgic dreams of the past, arrogant regional stereotypes, or scholarly arguments about central themes and unifying characteristics. Positive or negative, these images of the South keep us from seeing the people of the region with the fullness and empathy all people deserve. (Ayers 2005: 40)

White himself accounts for the film's approach thus:

> To some extent this film is not about 'the South', but about a romanticized version of the South seen through two oddly coupled anamorphic lenses, mine and the filmmaker's. The filmmaker's goal was not to reveal hard truths about a troubled region, but to express the sense of mystery that it and its by-products – principally in the form of folk music – inspired in them. They also made a decision to make a film strictly about the poor white South that they were fascinated with, saying that the larger issues of race were too much to tackle and not really the film they wanted to make. (Email correspondence, 4.4.2013)

The film recognises that plurality is important, hence the insights from insiders (the ruminations offered by prisoners, preachers, and bar-room drinkers, the self-declared 'backsliders') and outsiders (as with the various musicians). White hopes to provide some cultural context while also retaining something about the South as enigmatic. But he is reluctant to wear the mantle of ethnographer/anthropologist. His warning that the journey can only be undertaken if they drive the right car reveals his awareness that they should not look like outsiders while recognising that he will never be an insider. The film may be seen as a cross-cultural audio-visual exploration into Southern culture, but one that is aware of the subjectivity of such a process. This, then, is why White rejects the phrase 'southern state of mind' in favour of the South as an 'atmosphere', as something even more nebulous, impermanent, one that ultimately defies categorisation. As for the music, it provides us with a soundtrack to the route taken rather than the region itself.

Notes

1. Brett Sparks of the Handsome Family makes a similar point when he says 'the "South" that we are talking about, romanticizing, isn't really 'there' anymore. It's a ghost. I don't actually think the South we envision EVER existed' (email correspondence 23 September 2014).
2. http://www.fotomuseumdenhaag.nl/en/exhibitions/sally-mann (accessed 20 July 2014).

Works cited

Allen, Ray (2010). *Gone to the Country: The New Lost City Ramblers and the Folk Music Revival* (Urbana: University of Illinois Press).

Ayers, Edward L. (2005). *What Caused the Civil War? Reflections on the South and Southern History* (New York: W.W. Norton).

Barker, Hugh and Yuval Taylor (2007). *Faking It: The Quest for Authenticity in Popular Music* (London: Faber and Faber).

Baudrillard, Jean (1989). *America* (London: Verso).

Benjamin, Walter (1997). 'The translator's task' (tr. Steven Rendall), *TTR: Traduction, Terminologie, Redaction* 10(2), pp. 151–165.

Booth, Michael (2005). 'A certain South rises again', *Denver Post,* 10 July. http://www.denverpost.com/movies/ci_3089941 (accessed 20 July 2014).

Boym, Svetlana (2008). 'Ruinophilia: Appreciation of Ruins', http://monument-totransformation.org/atlas-of-transformation/html/r/ruinophilia/ruinophilia-appreciation-of-ruins-svetlana-boym.html (accessed 20 July 2014).

Cobb, James C. (1999). *Redefining Southern Culture: Mind and Identity in the Modern South* (Athens: University of Georgia Press).

Connell, John and Chris Gibson (2003). *Sound Tracks: Popular Music, Identity and Place* (London: Routledge).

Dubey, Madhu (2002). 'Postmodern geographies of the U.S. south', *Nepantla* 3(2), pp. 351–371.

Eco, Umberto (1998 [1976]). *Faith in Fakes. Travels in Hyperreality* (London: Vintage).

Felperin, Leslie (2003). 'Review: "Searching for the wrong-eyed Jesus" ', *Variety,* 17 December. http://variety.com/2003/film/reviews/searching-for-the-Wrong-eyed-jesus-1200537375/ (accessed 20 July 2014).

Filene, Benjamin (2004). 'O Brother, what next? Making sense of the folk fad', *Southern Cultures* 10(2), pp. 50–69.

Fink, Matt (2003). 'Sixteen horsepower – Olden', *Delusions of Adequacy*, August 4. http://www.adequacy.net/2003/08/sixteen-horsepower-olden/.

Fisher, Mark (2013). *Ghosts of My Life .Writings on Depression, Hauntology and Lost Futures* (Winchester: Zero Books).

Gill, Andy (2004). 'Album: Jim White', *The Independent*, 16 April, http://www.independent.co.uk/arts-entertainment/music/reviews/album-jim-white-6171062.html (accessed 20 July 2014).

Griffin, Larry J (2006). 'The American south and the self', *Southern Cultures* 12(3), pp. 6–28.

Grønstad, Asbjørn (2005). 'Wonders of the invisible world: The handsome family and the 'topographical uncanny', *Chapter & Verse. A Journal of Popular Music and Literature Studies*, Spring, http://www.popmatters.com/chapter/Issue3/gronstad.html.

Hutson, Cecil, K (1993). 'Cotton pickin', Hillbillies and Rednecks: An analysis of black Oak Arkansas and the perpetual stereotyping of the rural south', *Popular Music and Society* 17(4), pp. 47–62.

Knudsen, Britta Timm and Anne Marit Waade (2010). 'Performative authenticity in tourism and spatial experience: Rethinking the relations between travel, place and emotion', in Britta Timm Knudsen and Anne Marit Waade (eds), *Re-investing Authenticity. Tourism, Place and Emotions* (Bristol: Channel View Publications), pp. 1–21.

Lomax, John Avery and Harold William Thompson (1994 [1934]). *American Ballads and Folk Songs* (New York: Macmillan, Dover Books).

MacCannell, Dean (1973). 'Staged authenticity: Arrangements of social space in tourist settings', *American Journal of Sociology* 79(3), pp. 589–603.

Marcus, Greil (1997). *Invisible Republic. Bob Dylan's Basement Tapes* (London: Picador).

Moore, Allan (2002). 'Authenticity as Authentication', *Popular Music* 21(2), pp. 209–223.

O'Brien, Michael (2007). *Placing the South* (Jackson: University Press of Mississippi).

Orvell, Miles (1989). *The Real Thing. Imitation and Authenticity in American Culture, 1880–1940* (Chapel Hill: University of North Carolina Press).

Paz, Octavio (1992). 'Translation: Literature and Letters' (trans. Irene del Corral), in Rainer Schulte and John Biguenet (eds), *Theories of Translation. An Anthology of Essays from Dryden to Derrida* (Chicago: University of Chicago Press), pp. 152–162.

Petridis, Alexis (2014). 'Neil young: A letter home review – A gloriously gloomy album of lo-fi covers', *The Guardian* 22 May, http://www.theguardian.com/music/2014/may/22/neil-young-a-letter-home-review (accessed 20 July 2014).

Reed, John Shelton (1972). *The Enduring South. Subcultural Persistence in Mass Society* (Lexington, Mass.: Lexington Books).

Richards, Keith (2010). *Life* (London: Weidenfeld and Nicolson).

Roberts, Lisa C. (1997). From knowledge to narrative. Educators and the changing museum.(Washington, DC: Smithsonian Institution Press).

Smith, Barnaby (2010). 'Johnny Dowd' http://landlockedbluesblog.wordpress.com/2010/04/15/johnny-dowd/ (15 April accessed 20 July 2014).

Smith, Lindzee (1990). 'Nick cave', *BOMB* 31 (Spring) http://bombmagazine.org/article/1313/nick-cave (accessed 20 July 2014).

Sweetman, Simon (2014). 'Neil young's letter home', *blogonthetracks*, 4 June, http://www.stuff.co.nz/entertainment/blogs/blog-on-the-tracks/10117006/Neil-Youngs-Letter-Home (accessed 20 July 2014)

Urry, John and Jonas Larsen (2011). *The Tourist Gaze 3.0* (London: Sage).

Yaffe, David (2009). 'Bob Dylan and the Anglo-American tradition', in Kevin J. H. Dettmar (ed.), *The Cambridge Companion to Bob Dylan* (Cambridge: CUP), pp. 15–28

10
Emotional Places: The Role of Affect in the Relocation of Mancunian Melancholia

Georgina Gregory

> For two days in April, fans of a disbanded Mancunian pop group and its forgotten frontman ... picked over U.K. bootlegs and danced to *Hairdresser on Fire* like dehydrated Helen Kellers, which is how people at Smiths conventions are supposed to behave. Yet these fans are not the glowing white semi-goths you'd expect to encounter; this scene looks like a 1958 sock hop in Mexico City.
>
> (Klosterman 2006: 49)

The scene in Los Angeles described above, could hardly be further from the music's place of origin in the north of England where, due to the autobiographical character of much of his song writing, the music of the Mancunian songwriter Morrissey features multiple references to localities drawn from his native city. Morrissey rose to fame during the 1980s with his band the Smiths and subsequently pursued a solo career. Despite living for many years in Los Angeles and Rome, he is forever conflated with the city of Manchester. Indeed Morrissey's association with Manchester, referred to by O'Hagan (2007) as 'poetic provincialism', exemplifies the use of location as a source of identification. In addition to numerous allusions within song lyrics, the Smiths' record cover art also included signifiers relating to Manchester and northern England.

Somewhat surprisingly in view of the apparent cultural specificity of his songs and associated visual ephemera, Mexican fans in California express deep admiration for the work of the Smiths. More particularly as Arellano (2008) and Devereux, Dillane, and Power (2011) have noted, the band's erstwhile lead singer Morrissey, is embraced as an honorary member of the Mexican-American community in Los Angeles, a phenomenon documented by Aitch (2005) and Klosterman (2006). The growing number of Smiths/Morrissey fan nights and the popularity of the Mexican Smiths tribute band the Sweet and Tender Hooligans, are a testimony to his US Latino fan following. Morrissey's close connections with Manchester, the abstruse character of his lyrics, and the artist's regionally inflected accent make him an unlikely candidate for such adulation because on the surface the subject matter of his songs seems tangential to the day-to-day experience of life of Mexicans in California. His success with the Mexican community is even more puzzling in view of the fact that the high point of his career occurred some 30 years ago, long before many of these fans were born.

In order to understand why the 'Mancunian Miserabilist' has been appropriated as a Mexican idol and before providing any explanations for his relocation, I will begin by looking at how place informs the character of music more generally. Consideration will also be given to how references to specific locations can impinge on or facilitate reception and relocation of pop songs. Efforts to standardise and customise popular music texts by ironing out regional references are examined in the light of the shift towards the global marketing of popular music. Moreover, as lyrics play a significant role in the construction of popular music texts I evaluate the impact of language in the process of translation and relocation. Finally, I will show how in Morrissey's case, it is his ability to convey feeling and emotion which enables him to transcend geographical, temporal, and language barriers, allowing his fans to make their own interpretations of the music. In doing so, I aim to highlight the significance of embodiment, an aspect sometimes overlooked in studies of the relocation of music from place to place.

Popular music and place

Of course, Morrissey's use of his home town as a source of inspiration is by no means unique. The omnipresence of such allusions and the centrality of geography to popular music led Nick Coleman to observe that 'pop is better on cities than anything else' (Coleman 1995: 1). A number of academic studies substantiate his view, acknowledging the role of location in mediating the production as well as the consumption of the medium. For example Whiteley, Bennett, and Hawkins (2005) reflect on how the dynamics of economic and political circumstances have an impact on the places where music is experienced, produced, and consumed. The profound influence of local people and places on Morrissey's repertoire matches the observations of Cohen, (1991), Finnegan (2007), and Shank (1994) who see a close association between music and regional identity. Moreover, despite his identity as the archetypal northerner, he has also been described by (Bracewell 2002) as a quintessentially British persona, embodying Applegate and Potter (2002), Morra (2013), and Washabaugh's (2012) assertion that popular music articulates and expresses national as well as local identities.

Popular music's environmental determinism further extends to the rock and roll heritage industry where birthplaces, homes, and key landmarks in the lives of famous artists are fetishised. As a result, towns, buildings, streets, and monuments are imbued with significance and to fans they possess the mysterious aura of those who touched upon them however fleetingly. In this sense, they realise Bakhtin's observation that in the production of art 'temporal and spatial determinations are inseparable from one another, and always coloured by emotions and values' (Bakhtin 2002: 243). Hence the unusual rooftop setting of the Beatles' final concert in London is now a central pillar in the mythology surrounding the tragedy of their break up. In a similar spirit, despite the fact Johnny Marr and Morrissey now reside in the US and neither has lived in the city for decades, Smiths' fans regularly cycle around Manchester landmarks, stopping at key points across the city as a way of paying homage to the defunct ensemble.[1]

Cultural pessimists would no doubt attribute the omnipresence of all these spatial references in texts and practices to the industry's need to simulate an element of individuality. In Adorno's view, attempts to differentiate one popular music text from another offer only pseudo-individualism since he believed that, 'The whole structure of popular music is standardised, even where the attempt is made to circumvent standardization. Standardization extends from the most general features to the most specific ones' (Adorno 1941: 17). When we note Fiske's observation regarding, 'The failure rates for new products, whether primarily cultural, such as movies or records [which] are as high as 80–90 per cent despite extensive advertising' (Fiske 2006: 5), it is clear that any attempts to mitigate against such failure are welcome. Hence a steady stream of readily absorbed and easily recognisable imagery help to make a song stand out in a crowded market place. While desire, love, and loss are common themes, references to places real or imaginary draw forth emotion, anchor meaning, and create easy points of identification.

Adorno's cynicism would no doubt extend to the industry's efforts to secure attachments to artists on a regional basis, a practice designed to secure a local fan following. In the UK where the conflation of artist and region is most commonly presented within a north/south regional binary, the north invariably played the role of cultural underdog, a position the Smiths adopted throughout the early years of their career. In this way, they could be presented as different from southerners thus tapping into the English national obsession with social class. Local accents are also used to strengthen and enforce regional differences as witnessed in the 1960s when the distinctive dialect of Merseybeat artists provided a refreshing contrast to the overwhelming presence of 'received pronunciation' across contemporary British media platforms. For Alan Bennett the Liverpool accent acts as an 'essential medium for the projection and representation of the local micro-culture, the "scouse" blend of truculent defiance, collective solidarity, scallywaggery and fatalist humour which sets Liverpool and its inhabitants apart' (Bennett quoted in Belchem 2000: 33). Fortunately from the perspective of relocation beyond the UK, their accent was perceived as charming and somewhat exotic. Indeed at a US press conference, when asked to explain his strange English accent George Harrison allegedly said 'It's not English, it's Liverpudlian' (quoted in Leonard and Strachan 2010: 31).

Place in a global context

Aligning an artist with a specific location aids in many aspects of the packaging and promotion of popular music products, whether in genre categorisation or the planned exploitation of regional loyalties but the practice is not problem-free. In a global market, excessive emphasis on locality can have undesirable effects on sales, an outcome the industry seeks to eradicate. In Morrissey's case, the refusal to tone down his local accent undoubtedly helped to bolster his outsider status and cult following in the UK. On the other hand, the Smiths' resolute northern-ness was less appreciated by the mainstream American audience. The group's modest success there would suggest that overt signifiers of local identity alienated potential fans, especially those who did not understand or approve of the nuances of dialect or the specificities of local references.

For precisely these reasons accents and other distinctive signs of local identity are nowadays under threat, mainly due to the commercial shift towards globalisation and the concept of the homogenous market. Since the 1980s, globalisation has been hailed by economists as the final stage of capitalism, promising massive growth in economies of scale in production, distribution, and marketing.[2] Central pillars of this geographical expansion are the concepts of product standardisation and customisation: critics are advised that any loss of individuality will be more than made up for by increased sales and cost reductions (Levitt 1983). In common with other commercial organisations aiming to make their products accessible to the widest possible audience, the music industry has responded to a perceived need for greater uniformity.[3] It is recognised that financial returns from an artist who is only successful within a regional or even a national context will be concomitantly limited. In the words of one global marketing professional.

> A positive customer response in the artist's immediate region may be cause for celebrating [and] sales on a national level may seem like the ultimate success [but] there is further opportunity, and it lies beyond your country's borders. The sheer numbers of potential customers worldwide, the variety of the markets that global commerce offers, and the increasing ease of doing business around the globe make thinking globally a national extension of any music marketing plan. (Lathrop 2013, Chapter 1)

These comments present thinking globally as the ideal but as Negus (1999: 101) points out, although the vision of a 'homogenous market' prevails, in practice it does not exist and relatively few artists are able to achieve global success. Products are quite rigidly categorised along genre lines, certain styles of music directed towards particular audiences and music company divisions do not necessarily communicate well with one another. This makes Morrissey's popularity with Mexican-Americans all the more surprising because, following the wisdom of genre categorisation, marketing experts would anticipate this audience would prefer either traditional Mexican music, or Latin rock and pop. Directing the Smiths or Morrissey's repertoire towards the Latin audience would be an act of unthinkable folly.

Lathrop does at least acknowledge the difficulties in attracting foreign audiences, warning would-be global marketers that 'taste in music is so subjective, so unpredictable, so variable, so based on the most idiosyncratic of perceptions and preferences, that precisely defining a likely-to-buy segment of the music-consuming population remains largely an elusive goal'. He goes on to propose that unpredictability should be ameliorated by efforts to customise or standardise products aimed at foreign audience, offering the example of, 'A song [which] may not work in a country because its lyrics express a sentiment that falls outside of local concerns [or] is considered abrasive'.[4] In such a case, the writer suggests changing the lyrics and reworking the instrumental if a foreign market is the target. Following this dictum, a number of artists including Morrissey and the Smiths, would have fallen by the wayside from the outset and no doubt their creative output would not have found its way onto the global marketing menu.

Language and translation

Those involved in sales and marketing, perceive the language of the song to be the biggest single obstacle to standardisation. The Anglo-American domination of the popular music economy has secured the hegemony of the English language and only a small proportion of songs featuring foreign languages have managed to break into the UK and US charts.[5] Through the process of translation, music recorded in English can be customised for foreign language audiences – nevertheless, as Susam-Saraeva (2008: 187) reminds us, because each language has its own natural rhythm some match more readily with English rhythmic patterns than others.[6] Any reconfiguration or change to the lyrics will have an impact on the overall meaning of the original and, as Sapir (1958: 69) observes: 'No two languages are ever sufficiently similar to be considered as representing the same social reality.... We see and hear and otherwise experience very largely as we do because the language habits of our community predispose certain choices of interpretation.' The impact of national ideology, culture and values on the reception of music is a further issue and Worbs (1963) warns that the content of a recording or performance can easily alienate a potential audience if it touches on material which offends or is misunderstood due to cultural dissimilarities.

Hence it is interesting to note that despite the aforementioned hurdles some songs travel, seemingly without effort, from one continent to another. Plenty of British and American pop acts have attracted an international following without necessarily resorting to translation. For instance during the 1990s the Backstreet Boys topped the charts in Spain, Sweden, Belgium, Malaysia and Hungary and the recent global success of One Direction provides a more contemporary exemplar.[7] The Swedish pop group Abba were another equally successful international group, selling an estimated 350 million records worldwide.[8] Similarly, songs recorded in languages other than English do on occasion, make it into the UK and US charts, a notable recent example being *Gangnam Style* (2013) by the South Korean artist Psy.

The shared qualities of this international repertoire include catchy melodies, driving rhythms, and the upbeat character of the songs. This suggests that inoffensive and bland content promote universal accessibility. While fans across the globe quickly embraced songs like *Love Me Do* (1962) and *I Want to Hold Your Hand* (1963), it is unlikely they would have warmed so readily to the Beatles, had their early repertoire included challenging and controversial material such as *Tomorrow Never Knows* (1966) and *Revolution 9* (1968). On a superficial level, infectious tunes and uncomplicated subject matter travel more readily than more taxing or troublesome content.

From Manchester to Mexican America via Morrissey

If the same industry logic is applied to the Smiths' repertoire, although Johnny Marr produced many uplifting and catchy melodies, Morrissey's early song writing appeared to do little to promote global reach. In the songs *Rusholme Ruffians* (1988) and *Miserable Lie* (1984) he alludes to obscure city suburbs and in equally parochial manner the album title *Strangeways Here We Come* (1987), refers to the city's Victorian prison. Likewise, the unsavoury content of *Suffer Little Children* (1984) draws on horrific crimes committed by local child murderers. When compared to the bland pop beloved by multinational record companies, his creative output is decidedly challenging, and it is not surprising that mainstream American failed to embrace Morrissey, or the subject matter of his songs en-masse. Paradoxically, while he could hardly be described as deliberately interpolating them and only offered translated versions of his music after the fact, Mexican-Americans find his work accessible.

According to Mitchell (1996: 8) their adulation should not come as such a surprise. He shows how in certain circumstances music expressing local concerns, can be co-opted for other purposes and assigned with new meanings to represent the interests of a group geographically and culturally remote from the original site production. The factors governing these appropriations resist standardisation. While record companies see foreign language translation as the best strategy for globalising pop music, research undertaken by North, Hargreaves, and Hargreaves (2004: 63) reveals that audiences are able to relate to music without strict adherence to a linguistic narrative. This is because although there are differences in how the mind processes language and music, both are capable of generating meaning as Fitch notes:

> Spoken language and music are both complex, culturally transmitted hierarchical systems based on sound. To a first approximation,

> language is meaningful (in the sense of being capable of conveying an unlimited number of propositions) and music is not. However, music does convey *something,* as illustrated by the fact that we can reliably map a piece of music onto non-sonic domains (like dance or emotion). (Fitch quoted in Cutler 2005: 395)

These observations help to explain why Morrissey's Latino fans who may not fully appreciate the manifest content of his songs, the characters, narratives and nuances, still engage with him and his music. Furthermore, it should be noted that the artist's performance style strongly influences engagement because, 'Making music involves not only the communication of musical sounds but is also characterised by a continuously changing and meaningful use of facial expressions, body movements, and hand gestures' (Thompson et al. 2005: 156).

Undoubtedly in Morrissey's case, his dual Anglo-Irish identity, an issue highlighted in the title of the album *Irish Blood, English Heart,* provides an additional source of identification for Mexican-Americans who experience a hybrid existence. In the title track lines 'I've been dreaming of a time when / To be English is not to be baneful / To be standing by the flag not feeling shameful / Racist or partial ...', he reflects on the experience of living with a hybrid identity, a dilemma they share with him. In the words of one fan:

> We knew Morrissey grew up in Northern England from Irish parents. That experience, I'm told, is not unlike the experience growing up Mexican-American. Both of those cultures are similar – the Catholic upbringing, the love for boxing and soccer, the close-knit families, the working-class parents. Going to public school and feeling like an outsider because you're not quite with the cool kids because you're from someplace else. So even though he doesn't necessarily sing about those things now we relate to that guy who had a similar upbringing to what we did.[9]

As a northerner of Irish heritage in the UK, a country where the south dominates in matters cultural, political, and economic, Morrissey is experienced in the social dynamics of subalternship, a position his Mexican fans relate to in California where Menchaca's (1995) research shows Anti-Mexican sentiment is widespread.

There are further parallels between his retro-aesthetic sensibilities and the Mexican love of vintage Hollywood film stars. The artist's record cover designs have featured images of cult stars of the Hollywood era and as Iain Aitch writes 'Mexican-American culture is imbued with a devotion to 'oldies' from the stage, screen, and stereogram, with the sounds of 1970s soul' (Aitch 2005). Thus it could be argued that Morrissey's camp and dramatic stage persona reflects the aesthetic repertoire of artists associated with Mexican popular cinema. Similarities also exist, between Morrissey and the canonised actor/singer Eulalio González (alias El Piporro), famed for his emotional and melancholic ballads during the 1950s, in which he sang from the perspective of outsider, about his conflicting Anglo-American social and cultural influences. It is easy enough to notice links between his and Morrissey's repertoire, social position, and cultural influences.

Equally, one could point to another famous Mexican, the closet homosexual film star Ramón Novarro who rose to fame during the 1930s and like Morrissey, experienced a Roman Catholic upbringing which he found difficult to reconcile with his sexuality. While Morrissey has never admitted to being gay and sometimes claimed to be celibate or bisexual, he recently revealed that he had experienced a long-term intimate relationship with a man (Morrissey 2013: 273).[10] Furthermore, Morrissey's effeminate stage presence and emotional style of singing resemble that of the singer Juan Gabriel who performed in Las Vegas during the 1970s replete with flowing scarves, glittering costumes, and exaggerated pompadour hairstyles. Moreover according to Arellano '[Morrissey's] trembling falsetto brings to mind the rich, sad voice of the cult hero of Mexican cinema, Pedro Infante' (Arellano 2008: 129). As the third of 15 children from a working-class background, Infante achieved iconic stature among Mexicans, particularly in the period since his death as witnessed by the presence of admirers at his shrine and the number of musical tributes paying homage to him. In many respects, the level of adulation expressed towards Infante matches that directed towards Morrissey who is also perceived by many fans as an iconic working-class hero. Drawing on this rich repertoire of shared signifiers, fans in Los Angeles enjoy paying homage to both Morrissey and the icons of Mexican culture at 'Morrissey-oke' (karaoke) nights where they fashion their hair in pompadour style and dress in rockabilly style.

Connections with Mexico via popular song

Although a surface reading would suggest that music made by a Mancunian is unlikely to resemble Mexican popular song, some observers note a convergence between the affected quality of Morrissey's performances, the doleful tone of his songs and 'canción ranchera', a form of music associated with the traditional, pre-industrial Mexican way of life. The ranchero repertoire, laden with longing and lamentation, has been described as melodramatic, sentimental, and amatory by Mendoza (1998: 28). A similar connection is noted by Arellano who identifies a connection between Mexican music's 'morbid fascination with getting your heart and dreams broken by others, usually in death' (Arellano 2008: 129) and Morrissey's preoccupation with melancholia. Thus the lines, 'And if a double decker bus crashes into us. To die by your side would be a heavenly way to die' from the Smiths' song *There Is a Light That Never Goes Out* (1986) resemble 'The day that they kill me may it be with five bullets and be close to you' from Cuco Sánchez's song *Cama de Piedra* (1958). This suggests that despite major cultural, geographical, and temporal differences, the music shares common elements which are sufficiently significant to overcome perceived differences between the texts.

While it easy enough to see some common ground in the examples cited, the connections only partially explain how an artist who sings about wanting to be run over by an archetypally British double-decker bus, and laments the dilemma of having a girlfriend in a coma, has become a Mexican cult hero. The process whereby shared sentiments are communicated and received is less well understood than the fact that this kind of appropriation regularly occurs regardless and even despite industry interventions. The key to understanding Morrissey's appeal lies less in the manifest content of his songs and more in how he sings them. As Gilbert points out: 'The tone of the speaking voice determines responses to it more surely and immediately than does the content of the words spoken' (Gilbert 2004). We know for example, when someone hears or says a word in a language they do not understand, the brain still processes that word and from a psycho-linguistic point of view, this indicates that words create relationships with things, people, and places. Likewise, when we listen to someone talking or singing, we do not process what we hear in an entirely intellectual manner: instead we are moved by the way that a conversation or song make us feel.

This is particularly true of singing where the distinctive quality of the individual voice, referred to by Barthes as its 'grain', allows an artist to communicate a range of emotions which words alone may fail to do. Barthes' (1985) work on the voice is particularly useful because even though he was a philosopher of language he was keenly interested in modes of communication which transcend language, recognising the body as a crucial site in the production of meaning. Barthes was aware that together, gesture, vocal tone, and timbre influence meaning-making as much as words do. Frith concurs with this view arguing that our interpretation of pop songs is much less reliant on lyrics than it is on the noises which surround the words, reminding us how 'In daily life, the most directly intense statements of feeling involve just such noises: people gasp, moan, laugh [and] cry' (Frith 1983: 35). Perhaps this why for some people, a tribute artist never moves them in quite the same way that the original performer does. The original artist's unique utterances, stemming from deep within the body, their subtleties of gesture and posture, like our individual fingerprints, can never be fully replicated.

Affect and the relocation of emotion

Although it is overlooked in advice offered by marketing executives, the generation of feeling is such a fundamental aspect of pop's capacity to communicate. This is especially so as regards music with complex or challenging content. It also explains why a popular music artist with limited technical ability or a song with seemingly vacuous lyrics, can touch us deeply. The process by which these deep feelings are relocated, moving from one person or place to another, requires further investigation and I would suggest here that not all artists choose to exercise their communicative powers to the fullest extent. Indeed it would also be fair to say some may have limited ability to do so and Morrissey's aptitude in articulating uncomfortable feelings distinguishes him from a number of his contemporaries.

The process of communicating and eliciting feeling is further problematised by the fact that it is a two-way process. Theoretical studies of affectivity have traditionally focused on the interiority of the individual human subject but Deleuze throws further light by describing it in less subjective terms. While accepting that affect emerges from the individual he speaks of the experience as 'sensual intensities' which move through human bodies, referring to these intensities as 'becomings that go beyond those who live through them' (Deleuze 1995: 137). Hence affect is not a static entity arising from someone, rather it is in a continual state of flux, travelling to and fro from person to person. The performing artist can enable sensual intensities to surface within us, a process made easier when their physical presence creates a direct channel of communication with those present. Nevertheless, as Cumming notes, even when the artist is physically unavailable:

> It is obvious that musical sounds are not, in origin, an impersonal or accidental event, that they do not come to exist in the disembodied

> medium of a CD without the action of a performer's body, but technological intervention can induce a partial forgetfulness of this fact. (Cumming 2000: 21)

Morrissey's remarkable ability to activate his presence beyond the parameters of the recording is summed up by Simpson who describes him as 'one of the greatest pop lyricists – and probably the greatest-ever lyricist of desire – that has ever moaned' observing that 'he is fully present in his songs as few other artists are, in a way that fans of most other performers ... wouldn't tolerate for a moment' (Simpson 2003: 5).

Unlike artists who prefer to remain somewhat detached from their stage persona or reveal little about their personal life, Morrissey is prepared to offer access to some of his deeply uncomfortable feelings. Although he is renowned for using wit and irony, his songs are peppered with themes of self-loathing, loneliness, depression, hopelessness, and at times despair, all of which he explores unflinchingly. The experiences he articulates are common to all at different points in life, but few will ever be able to express them to the fullest extent. Morrissey's remarkable ability to communicate his personality and his feeling embodies the fundamental role of the artist – to draw forth a similar response from the audience: a point reiterated by Scherer and Zentner who note how:

> Both the stable identity (physical appearance, expression, reputation) and ability (technical and interpretative skills) of the performer, as well as transient performance-related variables ... (interpretation, concentration, motivation, mood, stage presence, audience contact, etc.) ... have a major impact on the perception and induction of emotion (Sherer and Zentner 2001: 361).

One way of understanding the global reach of Morrissey's communicative capabilities and how they override temporal and cultural barriers, can be found via reference to the 'Doctrine of the Affections', a body of theory embracing the idea that the aesthetic aim of music could and should be, to arouse specific emotions. In the words of a prominent proponent: 'Everything that occurs without praiseworthy Affections, is nothing, does nothing, is worth nothing' (Mattheson 1981: 146). Emanating from an era preceding the development of modern psychoanalytic theory, these thoughts can be seen as somewhat radical as Mattheson goes on to blame any failure to communicate the emotions on a composer's lack of insight into themselves. Such a charge could hardly be directed towards Morrissey who on the contrary, might be regarded as someone who suffers from a surfeit of self-analysis.

Mattheson also believed that emotions could be elicited through the use of certain procedures and devices which could create an involuntary response and although no standardised system was ever offered, some practical suggestions were proposed. For instance, the use of smaller intervals between notes should be used to signify melancholy.

> [Since] joy is an expression of our soul, thus it follows reasonably and naturally that I could best express this affect by large and expanded intervals. Whereas if one knows that sadness is a contraction of these subtle parts of our body, then it is easy to see that the small and smallest intervals are the most suitable for this passion. (Mattheson 1981: 104)

While not necessarily being the most technically accomplished singer, as a vocal artist Morrissey's strength lies primarily in his powers of emotional insinuation. In particular, he has a remarkable ability to emote melancholy for although the music which accompanies his vocal is often upbeat, light hearted even, his singing is more often minor mode inflected and his use of small intervals produces the quality of sadness or melancholia Mattheson outlines. The sadness is somewhat emphasised by the quavering uncertainty of his singing voice which adds a hesitance and lack of vigour more redolent of melancholy than cheer. Then if we consider the timbre of his baritone voice, it is characteristically darker than the male tenor colour and Morrissey's random use of yodels, yelps, sighs, and moans, all add texture and depth of feeling to his singing.

With his unique approach to vocal art, Morrissey exemplifies Mattheson's point that music can evoke universal sensations such as hope, despair, joy, and sorrow but I would argue that Morrissey's voice is also capable of articulating and evoking a subtler array of emotional experiences connected to his life experience as a disaffected and dislocated individual.

Hence his references to location are to some extent arbitrary. It is not necessary for Mexican fans to know that Whalley Range and Rusholme were down-at-heel Manchester suburbs during the 1980s in order to appreciate Morrissey's sense of isolation and alienation. Although he grew up there and his name is synonymous with Manchester, Morrissey's music could hardly be viewed as a celebration of the city anyway. Instead he positions himself as an outsider, offering the audience access to a universal location – the subject position of anyone who never felt entirely at home whether literally or existentially.

Thus rather than taking his audience on a literal journey to his Manchester hometown, he invites them to join him on a form of emotional tourism, using the primitive power of affect to transport them to places of pain, loss, and disaffection which are common to, and lie deep within all who believe that they do not fit in. Because his personal experience of living in Manchester was unhappy and problematic, these feelings colour the affectual residue of his repertoire. As the majority of Mexicans in the US are no doubt subjected to racial and cultural prejudice, they have no difficulty in identifying with his feelings of unhappiness. He elicits emotion from his audience through a form of emotional contagion, whereby expressing his own deep feelings he draws forth a similar response from listeners.

For Mexican-American fans, Morrissey's dilemma as a dislocated Anglo-Irish Mancunian perfectly matches their predicament. In common with others in a post-industrial society, where individuals are increasingly displaced one from another and from their extended families, as migrants they are forced into more impersonal and individualistic ways of living. Such experiences create a fertile environment for alternative means of establishing belonging. Within this scenario, Morrissey offers a form of intimacy which encourages others to engage with the intensity of his feelings of isolation, sadness, and alienation. Although never strictly designed for that purpose and without specific reference to Mexican or American locations, his minor-inflected melodies and emotionally charged vocalising encourage others to travel with him.

Conclusions

With the advent of digital media and the move towards globalisation, the representation of local and national identities through the medium of popular music is threatened by the growing practice of homogenisation. It is encouraging therefore to note audiences ignoring the homogenising strategies of the marketing executives who would prefer them to adopt particular subject positions and to make preferred readings and purchases. Their agency in determining how they will use popular music texts and practices shows that the culture industry's efforts are less well equipped to deal with the subtleties of subjective experience, non-verbal forms of communication or the elusive nature of affect since these qualities are less quantifiable or even tangible. Ultimately, although the industry may try to, it cannot regulate the meanings assigned to popular music's places and spaces or control their relocation. Relationships between space, identity, and popular music are becoming ever more complex and tenuous. Reassuringly for cultural optimists this means that efforts to standardise popular music will continue to be thwarted at the point of consumption. In conclusion, the relocation of Mancunian melancholia illustrates two important points. First, that 'Popular culture is always in process; its meanings can never be identified in a text, for texts are activated, or made meaningful, only in social relations and in intertextual relations'(Fiske 2006: 3). And second, 'The ideology of rock – the arguments about what records mean, what rock is for – has always been articulated more clearly by fans than by musicians (or businessmen)' (Frith 1983: 165). Morrissey's success in connecting with Mexicans in California illustrates the difficulties inherent in predicting how and where, in an increasingly fragmented society an artist may be relocated.

Notes

1. For information about the tour see: http://www.creativetourist.com/articles/festivals-and-events/manchester/manchester-weekender-wanna-be-in-my-gang-cycle-tour/
2. It should be noted that there is also a countertrend of emphasising the local or merging local and global, as the term 'glocalisation' suggests. With this rationale, products are developed and distributed globally, but are customised to accommodate the consumer in the local market.
3. Although there are few systematic studies of the industry's efforts to standardise, persuasive evidence of recent song writing practice in Wald's (2002) study of boy bands demonstrates how gender distinctions are eliminated from lyrics in order to reach the widest possible audience.
4. During the 1970s for instance the Osmond Brothers' record *Crazy Horses* was banned in South Africa as it was believed that the title alluded to heroin use (*The Guardian* 1999). Countless other pop songs have been banned for similar reasons – a recent example being recordings by both Lady Gaga and Beyoncé, banned by the Chinese government on the grounds of 'poor taste and vulgar content' (BBC 2011).
5. Isolated examples include the 1950s the Italian song 'Volare' which was a success in the US as was "Sailor" (1960) which featured the German language. 'Dominique' (French) and 'Sukayaki' (Japanese) reached the Billboard Hot 100 in 1963. More recently 'Gangnam Style' by the Korean artist Psy achieved international success.
6. The Beatles' efforts to offer German versions of their songs illustrate the difficulties of converting German, a language characterised by lengthy sentences and long words into English. In translation, the chorus of the song *Get Back* (in English 'Get back. Get back. Get back to where you once belonged') becomes 'Geh weg, geh raus. Geh raus nach deinem Haus', lines which are impossible to sing without modification, if the song's rhythm is to be maintained.

7. With the albums *Backstreet Boys* and *Backstreet's Back*.
8. Source: Rock and Roll Hall of Fame, http://rockhall.com/inductees/abba/bio/ (accessed 1 July, 2014).
9. Source: 'A Mexican Morrissey', *The Coffee Table Years,* 15 January 2012 http://coffeetableyears.wordpress.com/2012/01/15/a-mexican-morrissey/
10. Following the publication of his autobiography Morrissey wrote on his fan website *True to You,* 'Unfortunately, I am not homosexual. In technical fact, I am humasexual. I am attracted to humans. But, of course ... not many' (http://true-to-you.net/website) (visited August 12, 2013).

Works cited

Adorno, Theodore (1941). 'On popular music,' *Studies in Philosophy and Social Sciences* IX, 1, pp. 17–18.

Aitch, I. 'Mad about Morrissey', *The Guardian,* 25 March 2005 http://www.theguardian.com/film/2005/mar/25/popandrock (accessed 12 August 2014).

Applegate, Celia and Pamela Potter (2002). *Music and German National Identity* (Chicago: University of Chicago Press).

Arellano, Gustavo (2008). *Ask a Mexican* (New York: Simon and Schuster).

Bakhtin, Mikhail (2002).'Forms of time and of the chronotope in the novel', in Michael Holquist (ed.), *The Dialogic Imagination: Four Essays* (Austin: University of Texas Press), pp. 3–40.

Barthes, Roland (1985). *The Grain of the Voice: Interviews 1962–1980* (London: Cape).

Bennett, Andy in John Belchem (2009). *Merseypride: Essays in Liverpool Exceptionalism* (Liverpool: Liverpool University Press)

Bracewell, Michael (2002). *The Nineties: When Surface Was Depth* (London: Flamingo).

Cohen, Sara (1991). *Rock Culture in Liverpool: Popular Music in the Making* (Oxford: Clarendon Press).

Coleman, Nick (1995). 'The lights are much brighter there'. *The Independent,* http://www.independent.co.uk/arts-entertainment/the-lights-are-much-brighter-there-1615194.html (accessed 8 April 2015).

Cumming, Naomi (2000). *The Sonic Self: Musical Subjectivity and Signification* (Bloomington: Indiana University Press).

Deleuze, Gilles (1995). *Negotiations 1972–1990,* trans. Martin Joughin (London: Continuum).

Devereux, Eoin, Aileen Dillane and Martin Power (eds) (2011). *Morrissey: Fandom, Representations and Identities* (Chicago: The University of Chicago Press).

Frith, Simon (1983). *Sound Effects: Youth, Leisure, and the Politics of Rock* (London: Constable).

Finnegan, Ruth (2007). *The Hidden Musicians: Music-Making in an English Town* (Hanover: Wesleyan University Press).

Fiske, John (2006). *Reading the Popular* (London: Routledge).

Fitch, William Tecumseh (2005). 'Computation and cognition: Four distinctions and their implications', in Anne Cutler (ed.), *Twenty-First*

Century Psycholinguistics: Four Cornerstones (New York: Psychology Press), pp. 381–400.
Gilbert, Jeremy (2004). 'Signifying nothing: "Culture," "discourse" and the sociality of affect', *Culture Machine*, 6. http://www.culturemachine.net/index.php/cm/issue/view/1 (accessed 10 August 2014).
Klosterman, Chuck (2006). 'Viva Morrissey', in Chuck Klosterman (ed.), *IV: A Decade of Curious People and Dangerous Ideas* (New York: Scribner), pp. 48–56.
Lathrop, Tad (2013). *This Business of Global Music Marketing*, [Kindle DX version] retrieved from Amazon (New York: Billboard Books).
Leonard, Marion and Rob Strachan (2010). *The Beat Goes on: Liverpool, Popular Music and the Changing City* (Liverpool: Liverpool University Press).
Levitt, Theodore (1983). 'The globalization of markets', *Harvard Business Review*, May – June, 92.
Mendoza, Vincent (1998). La canción Mexicana: Ensayo de clasificación y antología (Mexico, D. F.: Fondo de Cultura Económica).
Mattheson, Johann (1981). *Johann Mattheson's Der vollkommene Capellmeister: A Revised Translation with Critical Commentary*, (trans. By Ernest C. Harriss) (Ann Arbor: UMI Research Press).
Menchaca, Martha (1995). *The Mexican Outsiders: A Community History of Marginalization and Discrimination in California* (Austin: University of Texas Press).
Mitchell Tony (1996). *Popular Music and Local Identity: Rock, Pop and Rap in Europe and Oceania* (Leicester University Press: London).
Morra, Irene (2013). *Britishness, Popular Music, and National Identity: The Making of Modern Britain* (London: Routledge).
Morrissey (2013). *Autobiography* (London: Penguin).
Negus, Keith (1999) *Music Genres and Corporate Culture* (London: Routledge).
North, Adrian and David Hargreaves and Jon Hargreaves (2004). 'The uses of music in everyday life', *Music Perception*, 22, pp. 63–99.
O'Hagan, Sean (2007). 'Morrissey – So much to answer for', *The Observer*, 6 May 2007, 11.
Sapir, Edward (1929). 'The status of linguistics as a science', in Sapir (ed.), *Culture, Language and Personality* (Berkeley: University of California Press).
Susam-Saraeva, Sebnem (2008). 'Translation and music: Changing perspectives, frameworks and significance', *The Translator*, 14, 2, pp. 187–200
Shank Barry (1994). *Dissonant Identities: The Rock 'n' Roll Scene in Austin, Texas* (Hanover: University Press of New England).
Sherer, Klaus and Marcel Zentner (2001). 'Emotional effects of music production rules', in Juslin, Patrick and John Sloboda (eds), *Music and Emotion: Theory and Research* (Oxford: Oxford University Press), pp. 361–392.
Simpson, Mark (2003). *Saint Morrissey* (New York: Touchstone).
Thompson, William, Phil Graham, and Frank Russo (2005). 'Seeing music performance: Visual influences on perception and experience', *Semiotica* 156, pp. 203–227.
Wald, Gayle (2002) 'I want it that way: Teenybopper music and the girling of boy bands', *Genders*, 35. http://www.genders.org/g35/g35_wald.html (accessed 1 September 2014).

Washabaugh, William (2012) *Flamenco Music and National Identity in Spain* (Aldershot: Ashgate).
Worbs, Hans (1963) *Der Schlager; Bestandsaufnahme, Analyse, Dokumentation, ein Leitfaden* (Bremen: C. Schünemann).
Whiteley, Sheila, Andy Bennett, and Stan Hawkins, (eds) (2005). *Music, Space and Place: Popular Music and Cultural Identity* (Aldershot: Ashgate).

11

Beauty Is Not the Word: Relocating Detroit in Eminem's Video *Beautiful*

Żaneta Jamrozik

Relocation is often used as a comparative category to describe moving from one place to another. But what if relocation occurs within one space? Or when the borders between cities, nations, and continents are blurred by political organisations, like the European Union or technology, like the Internet and mobile phones? French Marxist philosopher, Henri Lefebvre argues in his *Rhythmanalysis* (1991) that everyday activities, feelings and (pre)conceptions about space affect the space around us. Space according to him is organic and alive and, in turn, affects us back. 'Space has a pulse, and it palpitates, flows, and collides with other spaces' (Merrifield 2006: 105), what the French philosopher likened to sea waves. The coming and going of the waves on the sea is not a mere repetition, Lefebvre asserts, but the movement constituting a rhythm (Lefebvre 2004: 22). Rhythm, according to him, is not a perfect repetition but involves counter movements and layering of movements. It contains changes and errors; nevertheless it aligns together space, time, and the body of the experiencer, creating a 'pulse' of society and a sense of time and location. This linkage of space and time through the body of the experiencer explains why so many people find watching or listening to sea waves relaxing.

Layering is also a term frequently used in digital studies to discuss the role played by digital media, especially video and mp3, in the city space (Downey and McGuigan 1999; Laughey 2008; Lim 2014; Nunes 2006). The city space becomes layered space, full of signs, screens, ads, and people talking on phones, emailing, or listening to music. This kind of layered space Alessandro Aurigi and Fiorella De Cindio call 'augmented' (2008), while John Connell and Chris Gibson, reflecting on the effect digital media have on geographical borders, urge the use of 'more active terms' than 'global' and 'local' (2003: 17). Furthermore, the city space, being filtered through the body of the experiencer and his/her usage of mobile technology, is personalised, what Walter Benjamin described through his analysis of Charles Baudelaire's poems and Rosane Araujo stresses in the title of her book *The City is Me* (2013). In informational city, she argues, 'there is no "outside" ' and 'we can no longer see a city, nor enter or leave it' (ibid.: 6). Araujo's thesis resembles Benjamin's descriptions of Paris as a space rarely, if ever, remaining outdoors. The city rhythms and noises enter into Benjamin's room (Adorno and Benjamin 1994: 222) or are absorbed through his body. 'I had immediately absorbed Paris right down to my fingertips', writes Benjamin in the letter to Jula Radt from 1926 (Benjamin 1994: 292).

Linking Lefebvre's rhythm analysis with Benjamin's writings on the relation between the city and cinema, I shall discuss the medium of music video as a reflection on the layered and personalised character of modern city space. Referring to the above-mentioned call to redefine terms like 'global' and 'local' in a globalised world I will focus on hip-hop music video and its complex position, which, on one side, emphasises the places of origin of rappers, while on the other reaches out to global audience. In particular, I will analyse Eminem's video *Beautiful* (2009) in order to compare two kinds of spatial brandings: representation of Detroit as a layered city and Eminem (also known as Slim Shady and Marshall Mathers) as a branded, multiple persona. I am not interested in the extent to which Eminem actually produced the video, only how the video participates in creating the brand 'Eminem' through branding Detroit. In this approach, I follow Edward G. Armstrong's essay on Eminem's construction of authenticity by emphasising his roots in Detroit (2004) and Marcia Alesan Dawkins's assertion that 'the label "Eminem" does not refer cleanly to a real individual – it refers to a brand which paints Eminem as the only one that matters' (2010: 472). In what follows, I will argue that Eminem's self-promotion has been conducted by pointing to Detroit as his 'space of belonging'.

Walking as a method

Both Lefebvre and Benjamin present walking as a method for researching layered structure of the city and the rhythmanalyst/flâneur as an exemplary figure of a walker, who is simultaneously a poet and a researcher. Lefebvre foregrounds the body, writing that the rhythmanalyst must draw on 'his breathing, the circulation of his blood, the beatings of his heart and the delivery of his speech as landmarks' (2004: 21). For Benjamin, on the other hand, Charles Baudelaire embodies the qualities of the flâneur. The flâneur's 'awareness of his highly ambivalent situation – at once socially rebellious bohemian and producer of commodities for the literary market' – is what, according to Benjamin, accounts for his ability to teach intellectual producers (Buck-Morss 1989: 304).

Lefebvre describes walking similarly to sea waves: every next move is not the mere repetition but an addition to the previous one and accumulates in the body of the walker causing the feeling of tiredness, relaxation, or the sense of rhythmic co-operation with the surrounding space. Every passing tree is not a yet another tree but the tree in relation to the already passed and anticipated trees. According to the French philosopher, the experience of space and time is filtered through the body of the walker. The body works as a metronome, while rhythm serves as a measure enabling the walker both to experience and reflect on his/her experience through measuring it according to the rhythm of things which already happened and will happen (Lefebvre 2004: 89). Rhythm allows this type of meta-reflection on experience because, according to Lefebvre, it relies on both repetition and becoming and puts the Same in relation to the Other (ibid.: 79).This 'intra-organizational network', as Araujo calls the relation Lefebvre draws between the city and the person, constitute 'new spatial logic' (2013: 25). Hence, whenever a person is listening to music, in the car or on a personal stereo, s/he supplements spatial and bodily rhythms and changes the experience of space (Bull 2000, 2013; DeNora 2004). Walter Benjamin is often considered as a pioneer of researching this kind of relation between the city and technology. Like Lefebvre, the German philosopher is interested in strolling through the cityscape. However, unlike him, Benjamin stresses the fact that walking is a social act of 'taking up a position' or, to quote Madonna, 'striking a pose'. Walking through the city for Benjamin was a way to promote oneself, a tactic the philosopher advised to writers and journalists, and which today is, as it seems, partly replaced by Internet blogging. Walking for Benjamin amounts to going out onto the market, while the flâneur marks the increasingly blurred border between workers and the intelligentsia. 'He stands still on the thresholds', writes Benjamin, and 'seeks refuge in the crowd' in order to hide his in-between position. The flâneur marks then the 'intermediate stage' of relying on both patrons and the market. He is also a master of self-branding as he 'sets foot in the marketplace – ostensibly to look around but in truth to find a buyer' (Benjamin 2002: 10).

For both Benjamin and Lefebvre walking through the city is also an act of accessing history accumulated in the city space through enacting memory or habit learnt in the past. Lefebvre believes that every historical period has its own rhythm that governs the way people perform and experience things. It can be observed for example in the rhythm of speaking in historical dramas, like *Downton Abbey* (2010), in which people utter words significantly slower in comparison to today's standards. Everyday life and daily occurrence imply rhythm, which should be studied through learning from ordinary activities and repetitions of the era. The French philosopher calls for an examination of urban life in terms of regularity and repetition as rhythmanalysis, a method in which the researcher starts from his/her own body and its rhythmic organisation and then attempts to analyse in the same manner the social space surrounding him/her. 'To listen to one's own body is necessary to appreciate external rhythms', argues Lefebvre (2004: 19). The goal of Lefebvre's rhythmanalysis is to localise the subject in a greater whole of rhythms, what can be explained by quoting another French philosopher of everyday life, Jean-François Augoyard, who claimed that through everyday repetitions we are able to see that 'beyond the I, there is undoubtedly a *we*, which expresses itself in everydayness' (2007: 5).

Benjamin's approach to history, on the other hand, is through 'trash' – through the discarded materials that crowd the everyday. 'In this everyday material world different temporalities exist side by side: the latest version alongside last year's model', explains Ben Highmore. This layered structure of everyday debris inspired Benjamin to see history as a constellation rather than a line (Highmore 2002: 33). A linear representation of time and history, according to Benjamin and Lefebvre, does not fit into everyday life and everyday inhabitation of space, which relies on habit, repetition, and accumulation rather than progression. Hence they both refuse to see history as an 'after-one-another' single rhythm.

Reassuming, two people walking alongside each other but listening to different music through headphones, or accessing different information from Google maps, are, according to both Benjamin and Lefebvre, actually walking and experiencing two different spaces and learn about two different histories of space. If one of them searches for the nearest bus station and the other looks for a nice place to eat, the former will get only the map with the shortest route to the bus station, while the latter comes up with the whole list of the restaurants in the area. 'The information becomes a part of space' (Frith 2012: 132) and applications like Google maps remind us about it everyday.

Relocating the hip-hop sublime

In hip-hop music, spatial branding is a necessary part of rappers' self-branding and the link between the individual and the social is as important as in Lefebvre's and Benjamin's theories. Hip-hop culture acknowledges a paternal continuity between the space and the person, as can be seen on the cover of Nas's *Illmatic* (1994) where the artist's childhood photo is layered on to the one representing the Queensbridge Project, where he grew up. The 'hood' educates and, when needed, punishes. In order to gain credibility, rappers have to prove that they experienced this kind of schoolin'. Spatial roots, Steven Graves argues, are even more important in the case of white rappers and he points to Eminem as the best example (2009: 252).

Adam Krims calls the effect of the intentionally shocking video representations of life in urban ghettos, like the one present in NWA's video *Straight Outta Compton* (1988), the 'hip-hop sublime'. He argues that the hip-hop sublime is directed to people, who did not experience everyday ghetto life themselves and can access it only as a form of spectacle (2002). Krims suggests that these representations aim to picture inner city life as unrepresentable for anyone, who has not experienced it. The hip-hop sublime then reaffirms the dictum of the hood schoolin' saying that for those, who had not lived through it the ghetto is beyond comprehension.

The 'hip-hop sublime' is also a way of making money out of poverty in which poverty, paradoxically, 'becomes a source of surplus value' (ibid.: 65). The 'hip-hop sublime' returns then to the question of money, which is one of the main themes of hip-hop (Peterson 2006). The theme is summarised well in Nas's claim 'I'm out for dead presidents to represent me' (from his *The World is Yours* song), manifesting the artist's understanding that the only way to being poor and get represented nationally is to get rich and accumulate 'dead presidents' of dollar bills. There is then no way out of capitalism and the only critics of the system can come from inside the system. Poverty as a source of surplus value is also an important issue in the case of Eminem's *Beautiful,* representing some of the well-known post-industrial ruins of Detroit. The video could be easily dismissed as a hypocritical way of making money out of the city's wide recognition throughout popular culture as a place riven by economic depression and an iconic architectural degradation for the rapper, who no longer has to live in a trailer park but in Rochester Hills – a town outside Detroit where 82% of residents are white and unemployment is low. In his relocation, Eminem follows other famous rappers like Ice Cube or his mentor Dr Dre reminding that, against some popular opinions, hip-hop artists always perceived making money as a kind of public confirmation of their skills (Rose 1994: 56).

The hip-hop sublime as a way of creating surplus value from something that is supposed to have no value (the ghetto space of poverty) also supports Lefebvre's thesis that 'today, space as a whole enters into production' (2005: 155). Space becomes a product not because of its properties or localisation but as an image, or a brand. Space, according to Lefebvre, loses its use value on the account of stressing the exchange value. Space becomes a currency that can get exchanged into something else producing surplus value in the process, for example by authenticating Eminem's position as a rapper coming from Detroit.

Detroit is often channelled through the popular 'ruin porn' aesthetic: films or photographs emphasising Detroit's ruins as beautiful pieces of art and establishing an opposition between 'artistically' desolated buildings and those parts of the city that still remain a home for thousands of people. By establishing this opposition between Detroit as an icon of decline and Detroit as a living city, 'ruin porn' works as a political discourse that inclines the viewers to see Detroit's past represented by ruins as only aesthetically interesting, hence depoliticised. The ruins' beauty in 'ruin porn' works as Zygmunt Bauman's 'aesthetic of control' in which beauty becomes a way to get rid of any fear of danger, guilty conscience, or apprehension of shame and enable the viewers to experience Detroit 'aesthetically' without thinking of the political reasons for the city's decline (Bauman 1993: 6). The photographic representations of ruins usually try to catch the stillness and eternity of ruins, hence restrain from showing either people or movement within the frame. 'Given Detroit's long and ugly history of racial antagonism and the fact that it is at present nearly 90% African-American, it is possible', argues Nate Millington, 'to see "ruin porn" aesthetic as yet another attempt to erase African-American Detroiters from the landscape and reduce the representation of Detroit in media to the "city of ruins" ' (Millington 2013: 290).

It is important to think 'how a particular event or process (such as education, work, and poverty) occurs as a series of places' (Sack 1997: 255). Sack's remark returns when one reads headlines, like *Russia: One big Detroit* (Nemtsova 2013) or others asking 'why the future of Detroit is your future, too' (Ringen 2012: 33), as if Detroit was no more than a symbolic, futuristic, fictional city from the *Robocop* movies. As many writing on Detroit already pointed out, the city is caught between notions of decline and the promise of rebirth, what functions as a marketing strategy rather than a reflection on the city's past, present, and possible future (Kramer 2002, Longo 2006). Robert David Sack, similarly to Lefebvre, sees a danger in this kind of thinking that reduces the experience of the city to a slogan and sees its history as a straight line going in one direction: dystopic or utopic future. He reminds that 'landscapes are sustained through the rhythms and routines of everyday life' (Sack 1997: 8) and that past, present, and future intermingle in these everyday rhythms.

The chorus of the song *Beautiful* opposes the normative function of beauty as Eminem reminds his listeners: 'don't let 'em say you ain't beautiful', while the video for the song manifests a shift from the hip-hop sublime as a spectacle of shocking incomprehensibility to a more traditional version of sublime as a sensual (rather than intellectual) appropriation of beauty. In the case of the video it is a sensual appropriation of the space of Detroit. *Beautiful* is often mentioned as a track thematically different from other songs from the *Relapse* album. '*Beautiful* aside', noted Anthony Bozza, 'there's a real serial killer theme running through this album' (Bozza 2009). *Beautiful*'s lyrics tackle different subjects from zombies, leg and armless bodies or 60 sluts dying from asphyxia after sipping 'piss through a Christopher Reeves sippy cup' (all from the *Underground* track). However, the bodily sensations coming in excess on other tracks from *Relapse*, manifested by overflow of bodily fluids, like blood and urine or saliva-dripping zombies are also present in the *Beautiful* video through the appropriation of Detroit not as a metaphor of decline but as a sensuous, living space with which one can interact and even confess to. Place as Lefebvre suggests, can be depicted, performed, and sensed through its ensemble of normative and counter rhythms (2002: 264). The shift of the hip-hop sublime from the tactic of shock to sensual appropriation of the city space manifested in Eminem's video offers an alternative conceptualisation of Detroit, one which seeks to part with utopia and dystopia understood as grand narratives and formulate new, more nuanced or polyrhythmic, to refer to Lefebvre, means of thinking the city.

Detroit-ing

'The information I bring together in this book reflects my effort to move out of a fixed space called Detroit', writes Jeff Rice at the beginning of his *Digital Detroit* (2012: 13). Mobility, conceived by Lefebvre as the core of his rhythmanalysis is, as Rice's remark also suggests, a self-reflection over one's own role in creating space. As a self-reflection onto many layers of the space, Lefebvre's 'inner' mobility opposes capitalist mobility of constantly moving forward, beginning anew as manifested in the dictum of the American Dream (Keller 2005: 23–49) and reducing multidimensional space to only one dimension and one category, the future.

Emphasising one interpretation of the city in order to obscure another has been, Thomas J. Sugrue points out, a dominant way to present Detroit to the national and international public from the early 1950s (1996: 5). 'It is a commonplace observation', writes Sugrue, 'that Detroit's urban crisis begun with the riot of 1967 and worsened with the inauguration of Coleman Young as Detroit's first black mayor in 1974' (ibid.: 270). According to him, however, Detroit was experiencing decreasing of manufacturing jobs and racial discrimination long before the year 1967 when it was called by *The New York Times* the 'first major Third World City' (ibid.). In July 1967, a week-long rebellion by mostly black people fed up with the situation finally brought national media attention and had a large television coverage together with the later Kernel Commission's meetings that followed in order to find an answer to how could it have happened in the model city of Detroit (Burns 2004). The city that in the 1965 promotional film *Detroit: City on the Move* was branding itself as 'model', 'harmonious', 'multicultural' and 'modern', while bidding for the organising of the 1968 Olympics, after the 1967 Rebellion became known, after an article in *Life* magazine, as the national centre of disorder (Galster 2012: 202) and put in the same bag as communism, nuclear destruction, and the Vietnam War. The name Detroit became the synonym for 'disorder' and 'anarchy', as exemplified later in films like *Robocop* (1987), directed by Paul Verhoeven or Martin Brest's *Beverly-Hills Cop* (1984). The sense that the 'threats – both tangible and ideological – to American cities must be contained lest social anarchy occur was exemplified in the public and private responses to the Detroit riot' (Burns 2004: 4).

Lefebvre perceived the production of space as a conflict of antagonistic forces that is often 'smoothed out' by the dominant group. He welcomed the conflict and the mobility it brings with it as creative forces showing people the fractures in official perspective of space and work by, what he called, 'exploding' of space. My next aim is to present the *Beautiful* video and the network of meanings it intermingles with by circulating on the Internet as the device able to direct our attention on the movement as conflict, privileging it over ability to tell a story or giving an explanation of the presented events. The favouring of movement in both the content of music videos and the way they are often watched on mobile devices, in-between doing something else differentiate music videos from cinema and makes them more intertwined in our everyday, increasingly mobile, life.

The discursive relocation of Detroit after the Rebellion is one type of relocation discussed in this chapter; the other is the relocation of music videos from television to the Internet. Before the latter move, music videos were often discussed in the context of the television network (Kinder 1984; Goodwin 1992), but now they are presented as part of the digital network (Freston and Bruno 2009, Edmond 2012, McKinley Jr. 2013, Petchers 2012, Vernallis 2013b). Music videos are 'shaped technically and culturally by the colonising institutions and narratives of the visual media', Jody Berland argued in 1993 (Berland 1993: 21). This fact still holds true in 2014. However, because of the online relocation of music videos and their positioning next to other media content like news, commercials, or political campaign videos, music videos cease to be merely 'shaped by' and become also 'shapers' of public opinion or, as Marwan Kraidy put it, catalysts of public discourse (Kraidy 2012).

Music videos used to be conceived as commercials for a song or a performer. 'Videos are just commercials' (2004: 81), argued Carol Vernallis in 2004. In her 2013 article *Music Video's Second Aesthetic?* (2013a), she changed her mind and confessed: 'What is a music video? At one time we knew, but no longer; part of the change has to do with media contexts' (ibid.: 112). Marwan Kraidy suggests that thanks to the Internet music video became a catalyst of public discourse. Music videos, Kraidy argues, stimulate public discourse because of their popularity and capacity to conjure up several meanings and interpretations of a performer and put them into circulation with other meanings created by online platforms (2013: 271). Kraidy's essay suggests that the network of data activated by the video becomes more important than the video itself.

Coming and going (simultaneously)

Andrew Goodwin has noted that music videos rarely point to the lyrics, preferring to set a context or a mood corresponding to the song (Goodwin 1992: 22). Similarly, rhythm, according to Lefebvre, is a contextual matter, and rhythmanalysis is a method of drawing correspondences between spaces and events, rather than firmly discerning the types of relation between them. In this chapter I argue that music videos can be perceived as a tool of writing a kind of history that follows the rules of associations, 'grasps the constellations' rather than tells the sequence of events (Benjamin 2007: 263). The *Beautiful* video is one of those videos that is not interested in illustrating the lyrics literally but rather to create a greater, more nuanced or polyrhythmic context for their understanding. On its own, the song may seem to be yet another Eminem song about his depression, not least because it begins in this way:

> I'm just so fuckin' depressed, I just can't seem to get out this lump
> If I could just get over this hump
> But I need something to pull me out this dump

Later on, the song suggests that, once again, for Eminem this something to pull him out this dump is hip-hop, while the video brings Detroit into the picture, presenting the rapper and his own depression as a filter for the depression in the city. The video, especially at the beginning, restrains from doing what most hip-hop videos do, that is showing the artist's status in some community (Vernallis 2004: 75) as seen in other Eminem's videos set in Detroit (*Welcome 2 Detroit* with Trick Trick, for example). Instead, the video presents Eminem as a little confused and estranged while entering the space of Detroit's train station and works on suggesting relation between his brand image and the image of the city, what Jaimey Fisher points out as one of the characteristics of the flâneur's strolls (2005: 471). The video does not simply describe Detroit; rather, it uses the city as a way to define core elements of the Eminem brand and to brand the city through Eminem.

The video presents the rapper's walk through some of Detroit's best-known industrial ruins. Although the lyrics do not mention the name of the city, the video is able to establish the connection between Eminem and Detroit, by giving us the hint that somehow the origins of the lyrics are to be found in the experience of the devastated space of the city. The video aligns the personal history of Eminem's ups and downs and his wonderings about whether to quit his career as a rapper with the history of the city, which was once the heart of American industry and in a post-industrial era is forced to reinvent itself, reformulate its myth of a motor city. Indeed, the subtitles that open the video remind the viewer that 'In 1950 Michigan was 1 of 8 states in America that collectively produced 36% of the world's GNP. Detroit was the greatest manufacturing city in the world'. The content and font of the titles, looking as if manually composed and printed using hot metal matrices, remind one of Detroit as the proud capital of physical labour, workers, and trade unions. This laborious and glorious past is, however, contrasted with the labourless present. The video, through a subtle slow motion, emphasises the slow rhythm of passing time in today's Detroit and invites viewers to imagine how nowadays empty spaces must have looked and felt decades ago when they were full of people.

The music starts after the titles, and we can see Eminem's back as he enters one of Detroit's landmarks, the enormous Michigan Central Station (MCS), opened in 1913 and closed in 1988. Now, partly because of its devastation and partly thanks to the light coming through its many glassless windows, the station looks more like some kind of ruined Gothic cathedral. The gesture of entering brings with it both the person (Eminem) and the movement, distinguishing this shot from the representation of MCS on 'ruin porn' photographs, most famously on the cover of Yves Merchand and Romain Meffre's photo album *The Ruins of Detroit* (2010) (Figure 11.1).

Entering, Eminem passes a black, middle-aged lady, who, elegantly dressed and carrying a suitcase, looks as if she has just alighted from a train. The woman proceeds in the opposite direction to the rapper, rushing to the exit, and they both meet half way. This strange encounter of people, places, directions, and times, as the lady seems to be dressed in a fashion more typical of the 1950s than the present, reminds us that the building into which Eminem just entered is or rather was a train station. Now, however, the empty, ruined inside with the few remaining benches in the middle looks more like an empty church, devoid of any functional use (Figure 11.2).

The encounter shows two possible usages of the space of the MCS: for the lady, the station remains a station, and her behaviour there is similar to that in a typical Marc Auge's non-place (Auge 1995). She wishes to leave quickly and walks with a purpose. Eminem, on the other hand, behaves as if he finds himself in the MCS by an accident, just happens to be passing by, and decides to look at it. Since, as Vernallis

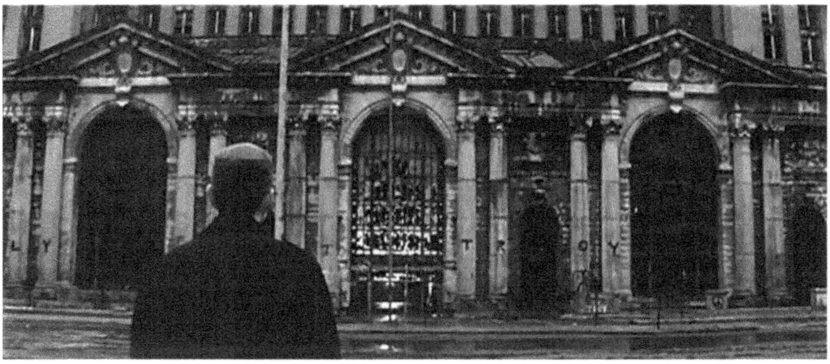

Figure 11.1 Eminem entering Michigan Central Station at the beginning of the *Beautiful* video

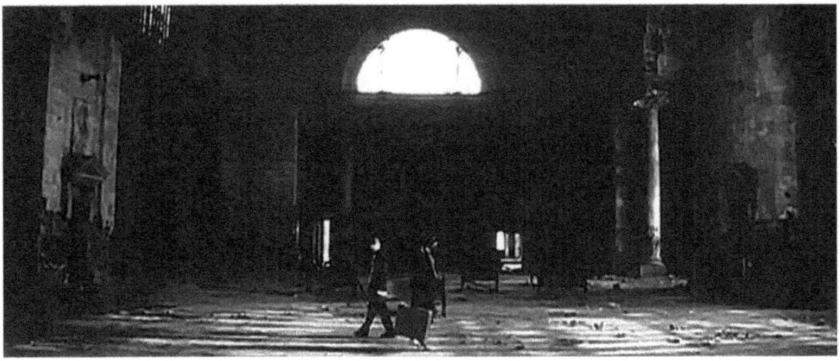

Figure 11.2 The encounter of people, places, directions, and times inside Michigan Central Station in the *Beautiful* video

argues, videos teach us how to dance to certain types of music, how to dress, and even how to 'feel about our bodies within a certain cultural epoch' (2004: 71), it is fair to compare Eminem's confusion to the one described by many Detroiters born after the city's heyday to whom its glamorous past of the 1950s is as alien as the futuristic Detroit from Paul Verhoeven's *Robocop* (1987) (Binelli 2012: 12). In this sense, the video teaches viewers this confusion and inspires them to imagine what it is like to live in Detroit's 'land of broken dreams'.

The MCS in *Beautiful* video is both familiar and alien and it resonates onto the rest of Detroit's buildings shown in the video and adds another layer to Eminem's brand persona – at once familiar and alien, from Detroit and not, since there are many Detroits. It remains the most important and the most featured building in the video. It opens and closes *Beautiful*, framing the video. It functions as a space of initiation, rehearsal, and training as we can see at one point Eminem picking up a microphone and rapping the lyrics 'But I need that spark to get psyched back up / In order for me to pick the mic back up' as if rehearsing and weighing his chances of going back to be a star rapper. The station serves then as a reflexive pathway, being something in-between the private sphere of the bedroom and the public sphere of the city. For Eminem, it becomes a kind of church in which he raps, confessing that he wonders whether to 'quit rap' and that he has some doubts as to whether he is still as good as he was. What is important for my analysis of layered space of the modern city is that this function of mediating between the private and the public, inside and outside the station gains through its initial positioning as both alien and familiar, being part of both the past and the present. Those two interpretations or visions of the building, coming from the lady and Eminem's reactions to it, exist simultaneously and even encounter each other in the moment when the rapper and the woman meet in the middle of the great hall of the station.

As already mentioned, the big, and largely glassless, windows of the MCS and numerous holes in its walls transmit small rays of sunlight that can make the station look almost cheerful and a second later really depressing. These rays of light function as, what Carol Vernallis calls, 'contagious' elements. 'Many music videos', writes Vernallis, 'work on the principle of contagion: an element in one of the strands seeps into another – it might be a color, a particular prop, a way of feeling or moving' (2004: 21). On occasions, the light in Eminem's video comes in strains so strong that they obscure rather than reveal the object, like in the scenes where the rapper is shown walking at night in Detroit. The light is always present and its presence only increases throughout the video up till the final moments showing Eminem at Detroit's Tiger Stadium. It is one of the features that move us throughout the video, make the space look more alive and dynamic and function as another layer of things – either revealing or obscuring what happens in a frame.

It is also light that most visibly divides the video into two parts: the first, beginning with Eminem entering Michigan Central Station and the second, starting from him visiting the demolition of Detroit's Tiger Stadium, where we can see more Detroit citizens. The parts sometimes intermingle with each other, but the difference of shots (darker shots of the ruins and sunnier shots of the stadium), keeps them recognisable. The shots taken outside the ruins show the people of Detroit, together with Eminem, looking into the camera, which is another change compared to the shots of the ruins. This way of looking at the camera is stressed as a kind of statement and, as one commentator on the rap genius website noted, 'it kinda gives you the feeling that Eminem is talking to "Detroit" the city' (FroDoDaGod). This is one possible interpretation. However, the video features similar looks from the city's citizens, which suggest first, a kind of unity of looking between them and Eminem and second, that the look is not so much directed onto the city as projected by the city onto the rest of the world (Figure 11.3).

The question of gaze is important in music video and differentiates it from cinema. 'Looking at someone', Benjamin observed, 'carries the implicit expectation that our look will be returned by the object of our

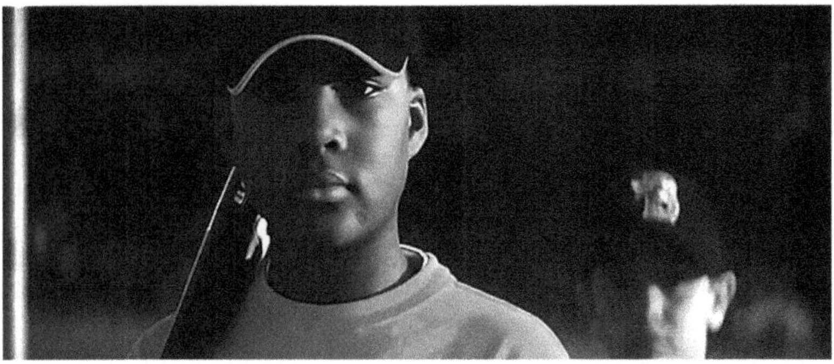

Figure 11.3 People of Detroit looking back at the viewers at the end of the *Beautiful* video

gaze' (2007: 188). Music videos answer this expectation by questioning the boundary between the space of the performer and the space of the viewer. The gaze of the performer in music video, unlike cinema, is often directed towards the viewers acknowledging their presence, which sets up a spatiotemporal relation between the spectator and the artist. Laura Frahm calls this kind of video 'photographic' (2010: 174), arguing that they often have the stillness more characteristic of photography than video and feature lots of close-ups. This kind of photographic stillness fills the last minute of *Beautiful*, as the video does not feature walking but focuses on the stillness of the gaze directed towards the camera by Eminem and the people of Detroit gathered around Tiger Stadium. The gaze seems also a kind of a challenge posed towards the viewers to remind them that by looking they are, like the flâneur, taking up the position towards the city and encourage them to notice their part in creating the image of Detroit. Marshall Berman in his analysis of the experience of modernity claims that 'we don't know how to use our modernism', and 'we have missed or broken the connection between our culture and our lives' (1988: 24). However 'mostly', he continues, we have 'lost the art of putting ourselves in the picture' and 'of recognizing ourselves as participants and protagonists in the art and thought of our time' (ibid.). *Beautiful* tries to reincarnate this art. The ending of the video shows the ability of the medium to include viewers 'into the picture' and make them conscious of their activity of watching and experiencing music. The ability to move from personal to public level of history is crucial for Lefebvre's theory of space and his conception of history (Kofman and Lebas 1996: 9). The history of Detroit as preserved in its space, considered the symbol of decay by many, in the video, filtered through Eminem's biography, is presented as a symbol of rebirth and recovery, a necessary grounding centre, and a place in the middle of private and public, a symbol of industry and a place to live.

Beautiful network

The *Beautiful* video and the film *8 Mile* (2002) are linked through their focus on Detroit. Both are not only set in the city, featuring some of its well-known buildings but also share the same iconography dominated by the greyish blue colour palette. The city is also featured on the cover of Eminem's next album, *Recovery* (2010) and in the Chrysler 200 commercial titled *Born of Fire* (2010), presenting the rapper driving the Chrysler through Detroit. All four show the more 'serious' side of Eminem and, if *My Name Is* or *The Real Slim Shady* introduced Eminem's 'id persona' Slim Shady, *Beautiful* and *8 Mile* offer more tender 'ego' side of the rapper as Marshall Mathers (Gilbey 2003). *Recovery*'s cover features Eminem sitting in what seems like a living room with a book in his hand and placed in front of a TV – a picture of ordinariness except for the fact that the living room is surrounded by a glass cube and behind it there is Detroit's Renaissance Centre, a group of skyscrapers which is one of the city's landmarks. The cover locates Eminem in the centre of Detroit but at the same time shows this centre as 'detached' or artificially constructed since the living room is not part of any home. The Renaissance Centre whose name suggests rebirth of Detroit is also an ambiguous symbol as the buildings were erected in the 1970s, not long after the 1967 Rebellion and were seen by many as forcing the 'rebirth' theme in order to gain financial advantage by attracting tourists and businessmen to the city and doing it by forgetting the past (Longo 2006). In the context of Eminem's career, *Recovery* was also an album made to overcome *Relapse*'s moderate success as the latter sold fewer copies than expected and Eminem himself expresses disappointment in his rhymes from *Relapse* a year later on his *Not Afraid* track.

Similar themes of come-back and rebirth are present in the Chrysler 200 commercial. As Nicholas Cook points out, commercials are too short to focus on the temporality of the experience of music. Therefore, if they use well-known music, they do that as a kind of iconic gesture, referring to the artist's background and image and transfer the values embodied in those onto the advertised product. Commercials 'made music mean', Cook argues, by materialising what is commonly associated with a given type of music. Instead of actually letting people experience music for themselves, commercials show how other people experience it or transpose its features (dynamism, boldness, etc.) onto the branded product and 'save time' (Cook 1994: 35). The Chrysler 200 advert uses Eminem's *Lose Yourself* from the *8 Mile* soundtrack as, Melissa Garlick, head of Chrysler advertising explains, as a well-known 'come-back' anthem (Bierly 2011) and grafts it onto images showing the rapper driving Chrysler in Detroit.

The commercial for Chrysler 200, often referred to as 'the Eminem ad', was shown for the first time during Super Bowl 2011's commercial break. It was regarded as the first spectacular manifestation of Chrysler's viability since the firm went bankrupt and was then bailed out by US government at the beginning of 2009. The journalist, Nancy Skinner on *America Live* (Fox News on 8 February 2011) began her programme stating: 'If there is buzz they want, they got it. Chrysler's big Super Bowl ad is sparking a major debate today. The auto-makers spending 9 million dollars on an ad featuring Detroit-native Eminem when they still owe American tax-payers billions of dollars for bailing them out. So, did you get your money's worth?'

Also, many observed that the ad is confusing in its inconsistency. In the off-screen narration, it refers to Detroit as a manufacturing centre while showing only service jobs like policeman, doorman, ballet dancer and Eminem himself. It is hard to tell what Chrysler is selling: Detroit, Eminem, the car or the Italian-American company's allegedly great position in the community. Generally, however, the common opinion is that it actually tries to sell a car by selling the city, as Bertel Shmitt summarises in the title of his article 'Chrysler has a city to sell to you' (Shmitt 2011). Chrysler uses Detroit as an orientation point for Americans to convince them to buy Chryslers because they are 'imported from Detroit' as the tagline says. However, for Detroiters it is more important where they are produced and, as the ad shows, it is not Detroit.

Born of Fire is an example of using what Kevin Lynch calls 'the image of the city' (1960), that is a central point according to which people orient themselves in their everyday lives. Lynch describes how people make meaning out of space by moving through it in their daily life, finding their way as they go. Cities, according to him, are at once dynamic constructs created 'on the go' and stable points of reference for creating one's identity. Chrysler, however, wishes to use Detroit's image as an industrial centre and connect it to Eminem and the theme of rebirth in order to organise American's car buying.

The image of the city, Kevin Lynch asserts, inspires a way of acting that is habitual and habits are imagined as connected with space like in mnemotechnics. The disappearance of manufacturing cities and the prominence of global cities as the centres of consumption rather than production are as Lefebvre observes, a highly unbalanced, monorhythmic vision of the world in which everything including the city is a product to consume (2003: 297). Global cities are often conceived as 'end points' that create a network of circulation of goods, while remaining static and unchanged. They are also used to diminish any thinking on the actual production that occurs in China or India. Chrysler's ad uses Detroit in this way, while at the same time branding it as a manufacturing city, a centre of production. It also forces the theme of rebirth describing Detroit as 'a town that's been to hell and back' and focuses on the future by obscuring the past. The example of Detroit as presented in Eminem's *Beautiful* video offers an alternative conception of the city. Detroit is presented in the video as opposing the monorhythmic agenda of consumption propagated by global cities by picturing Detroit as multi-layered space where the past, the present, and dreams of the future can exist next to each other. Thanks to this representation, the city and its rhythms work as the central or grounding point, not merely as a circulatory system for capital, people, and information. The video, presenting Eminem walking through the post-industrial spaces of Detroit, offers a conception of history as sedimented in a territory. It also presents Eminem turning to the city space in order to sort out his personal problems and finish the song.

Conclusions

In this chapter, I aimed to show in the example of Detroit and Eminem's video that the contradictions of space described by Lefebvre can nowadays be expressed not through cinema as in Benjamin's times but through music videos and the network of meanings they create circulating on the Internet. The ability to move from personal to public level of history, crucial for Lefebvre and Benjamin, organises Eminem's video *Beautiful*. Detroit's remaining between utopia and dystopia, always on the brink of recovery, reappears in Eminem's self-branding enabling him to map the differences and contradictions of Detroit's space onto his image of the white rapper and create a multiple Eminem-Slim Shady-Marshall Mathers persona. Unlike Chrysler's commercial that attempts to resolve the contradictions by presenting Detroit as recovered dystopia, the city that 'has been to hell and back', *Beautiful* pictures Detroit as the space 'in the middle', not the 'end point' of global production but the space of constant layering and relocation between the past, the present, and the future or notions such as utopia and dystopia.

Works cited

Araujo, Rosane (2013). *The City is Me* (Bristol: Intellect).

Augoyard, Jean-François (2007). *Step by Step: Everyday Walks in a French Urban Housing Project* (Minneapolis: University of Minnesota Press).

Adorno, Theodore and Benjamin, Walter (1994). *The Complete Correspondence 1928–1940* (New York: Blackwell).

Armstrong, Edward G. (2004). 'Eminem's Construction of Authenticity', *Popular Music and Society* 27(3), pp. 335–355.

Aurigi, Alessandro and De Cindio, Fiorella (2008). *Augmented Urban Spaces: Articulating the Physical and Electronic City* (London: Ashgate).

Augé, Marc (1995). *Non-places: Introduction to an Anthropology of Supermodernity* (London, New York: Verso).

Bauman, Zygmunt (1993). *Postmodern Ethics* (Oxford: Blackwell).

Benjamin, Walter (1994). *The Correspondence of Walter Benjamin, 1910–1940*, Gershom Scholem and Theodor W. Adorno (eds) (Chicago: The University of Chicago Press).

Benjamin, Walter (1999). *The Arcades Project* (New York: MIT Press).

Benjamin, Walter (2007). 'On Some Motifs in Baudelaire', in *Illuminations* (New York: Random House), pp. 155–195.

Berland, Jody (1993). 'Sound, image and social space: Music video and media reconstruction', in Simon Frith, Andrew Goodwin and Lawrence Grossberg (eds), *Sound and Vision: Music Video Reader* (London: Routledge).

Berman, Marshall (1988). *All That Is Solid Melts into Air: The Experience of Modernity* (New York: Penguin).

Bierly, Mandy (2011). 'Eminem's Chrysler Super Bowl Ad: Behind the scenes of one of Sunday's Best Commercials', *Inside TV*, http://insidetv.ew.com/2011/02/08/eminem-chrysler-super-bowl-commercial/

Binelli, Mark (2012). *Detroit is the Place to Be* (New York: Metropolitan Books).

Bozza, Anthony (2009). 'Being Eminem', *The Observer* 17 May 2009, http://www.theguardian.com/music/2009/may/17/eminem-urban-music-relapse

Buck-Morss Susan (1989). *The Dialectics of Seeing: Walter Benjamin and the Arcades Project,* (Cambridge, London: MIT Press).

Bull, Michael (2000). *Sounding out the City: Personal Stereos and the Management of Everyday Life* (New York: Berg).

Bull, Michael (2013). 'Remaking the urban: The audiovisual aesthetics of Ipod use', in John Richardson, Claudia Gorbman, and Carol Vernallis (eds), *The Oxford Handbook of New Audiovisual Aesthetics* (Oxford: Oxford University Press), pp. 628–644.

Burns, Andrea A. (2004). 'Waging cold war in a model city: The investigation of "subversive" influences in the 1967 Detroit Riot', *Michigan Historical Review* 30(1), pp. 3–30.

Connell, John and Gibson, Chris (2003). *Sound Tracks: Popular Music, Identity and Place* (New York: Routledge).

Cook, Nicholas (1994). 'Music and meaning in the commercials', *Popular Music* 13(1), pp. 27–40.

Dawkins, Marcia Alesan (2010). 'Close to the edge: The representational tactics of Eminem', *The Journal of Popular Culture* 43(3), pp. 463–485.

DeNora, Tia (2004). *Music in Everyday Life* (Cambridge: Cambridge University Press).

Downey, John and McGuigan, Jim (1999). *Technocities* (London: Sage).

Edmond, Maura (2012). 'Here we go again: Music videos after YouTube', *Television New Media* 20, pp. 1–16.

Fisher, Jaimey (2005). 'Wandering in/to the rubble-film: Filmic flanerie and exploded panorama after 1945', *The German Quarterly* 78(4), pp. 461–480.

Frahm, Laura (2010). 'Liquid cosmos: Movement and mediality in music video', in Henry Keazor and Thorsten Vubbena (eds), *Rewind, Play, Fast Forward: The Past, Present and Future of the Music Video* (New Brunswick and London: Transaction Publishers), pp. 155–179.

Freston, Tom and Bruno, Antony (2009). 'Music videos migrate online', *Billboard* 121(51), p. 112.

Frith, Jordan (2012). 'Splintered space: Hybrid spaces and differential mobility', *Mobilities* 7(1), pp. 131–149.

FroDoDaGod, rapgenius. http://rap.genius.com/Eminem-beautiful-lyrics#note-26551

Galster, George (2012). *Driving Detroit: The Quest for Respect in the Motor City* (Philadelphia: University of Pennsylvania Press).

Gilbey, Ryan (2003). 'In the ghetto', *Sight and Sound* 2, pp. 36–39.

Graves, Steven (2009). 'Hip-hop: A postmodern folk music', in Ola Johansson, Thomas L. Bell (ed.), *Sound, Society and the Geography of Popular Music* (London: Ashgate), pp. 245–260.

Goodwin, Andrew (1992). *Dancing in the Distraction Factory* (Minneapolis: University of Minnesota Press).

Highmore, Ben (2002). *Everyday Life and Cultural Theory* (New York: Routledge).

Keller, Florian (2005). *Andy Kaufman: Wrestling with the American Dream* (Minneapolis: the University of Minnesota Press).

Kinder, Marsha (1984). 'Music video and the spectator: Television, ideology and dream', *Film Quarterly* 38(1), pp. 2–15.

Kofman, Eleonore and Lebas, Elizabeth (1996). 'Lost in transposition – Time, space and the city', in Kofman, Eleonore and Lebas, Elizabeth (eds), *Henri Lefebvre: Writings on Cities* (Oxford: Blackwell), pp. 1–60.

Kraidy, Marwan M. (2012). 'Contention and circulation in the digital middle east: Music video as catalyst', *Television New Media* 14(4), pp. 271–285.

Kramer, Michael J. (2002). '"Can't forget the motor city": Creem magazine, rock music, Detroit identity, mass consumerism, and the counterculture', *Michigan Historical Review* 28(2), pp. 42–77.

Krims, Adam (2002). 'The hip-hop sublime as a form of commodification', in Regula Burckhardt Quereshi (ed.), *Music and Marx* (London: Routledge), pp. 63–78.

Laughey, Dan (2008). 'Music media in young people's everyday lives', in Jamie Sexton (ed.), *Music, Sound and Multimedia* (Manchester: Edinburgh University Press), pp. 172–188.

Lefebvre, Henri (2003). *The Urban Revolution* (Minneapolis: University of Minnesota Press).

Lefebvre, Henri (2004). *Rhythmanalysis: Space, Time and Everyday Life* (London: Continuum).

Lefebvre, Henri (2005). *Critique of Everyday Life Volume III: From Modernity to Modernism* (London: Verso).

Longo, Julie (2006). 'Remembering the renaissance city: Detroit's bicentennial homecoming festival and urban redevelopment', *Michigan Historical Review* 32(2), pp. 89–118.

Lim, Merlyna (2014). 'Seeing spatially: People, networks and movements in digital and urban spaces', *International Development Planning* 36(1), pp. 51–72.
Lynch, Kevin (1960). *The Image of the City* (Massachusetts: MIT Press).
McKinley, James (2013). 'Pop videos? I want my YouTube!', *New York Times* 162(56237), pp. C1–C2.
Merrifield, Andy (2006). *Henri Lefebvre: A Critical Introduction* (London: Routledge).
Millington, Nate (2013). 'Post-industrial imaginaries: Nature, representation and Ruin in Detroit', *International Journal of Urban and Regional Research* 37(1), pp. 279–96.
Nemtsova, Anna (2013). 'Russia: One big Detroit', *Newsweek Global* 161(44), p. 23.
Nunes, Mark (2006). *Cyberspaces of Everyday Life* (Minneapolis: University of Minnesota Press).
Petchers, Brian (2012). 'The Branding Power of Today's Music Videos', *Forbes* 952, pp. 66–69.
Peterson, James (2006). '"Dead Prezence": Money and mortal themes in hip-hop culture', *Callaloo* 29(3), pp. 895–909.
Rice, Jeff (2012). *Digital Detroit: Rhetoric and Space in the Age of the Network* (Philadelphia: Southern Illinois University).
Ringen, Jonathan (2012). 'Why the future of Detroit is your future, too', *Rolling Stone* 1166, p. 34.
Rose, Tricia (1994). *Black Noise: Rap Music and Black Culture in Contemporary America* (Hanover: Wesleyan University Press).
Sack, Robert David (1997). *Homo Geographicus A: Framework for Action, Awareness, and Moral Concern* (Baltimore: John Hopkins University Press).
Shmitt, Bertel (2011). 'Chrysler has a city to sell to you', *The Truth About Cars*. http://www.thetruthaboutcars.com/2011/02/chrysler-has-a-city-to-sell-to-you/
Sugrue, Thomas J. (1996). *The Origins of Urban Crisis: Race and Inequality in Postwar Detroit* (New York: Princeton University Press).
Vernallis, Carol (2004). *Experiencing Music Video* (New York: Columbia University Press).
Vernallis, Carol (2013a). 'Music video's second aesthetic?', in John Richardson, Claudia Gorbman, Carol Vernallis (eds), *The Oxford Handbook of New Audiovisual Aesthetics* (Oxford: Oxford University Press), pp. 437–465.
Vernallis, Carol (2013b). 'Audiovisual change: Viral web media and the Obama campaign', *Cinema Journal* 50(4), pp. 73–97.

Index

Printed and bound by CPI Group (UK) Ltd, Croydon, CR0 4YY